40 Advances in Polymer Science

Fortschritte der Hochpolymeren-Forschung

Luminescence

With Contributions by
E.V. Anufrieva, K.P. Ghiggino, Yu.Ya. Gotlib,
D. Phillips and A.J. Roberts

With 64 Figures

Springer-Verlag
Berlin Heidelberg New York 1981

no anal

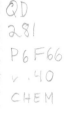

ISBN-3-540-10550-6 Springer-Verlag Berlin Heidelberg New York
ISBN-0-387-10550-6 Springer-Verlag New York Heidelberg Berlin

Library of Congress Catalog Card Number 61-642

Table of Contents

Investigation of Polymers in Solution
by Polarized Luminescence

Elizaveta V. Anufrieva, Yuli Ya. Gotlib

The Institute of High Molecular Compounds, Academy of Sciences of the USSR,
Leningrad 199004, USSR

In recent years, the method of polarized luminescence (PL) has become one of the most valuable methods for the study of macromolecules. It provides detailed information on micro-Brownian motion in polymer chains which is related to various features of the structures and conformational properties of macromolecules.

This review considers the use of PL in the investigation of macromolecules with covalently bonded luminescent markers (LM). This approach is very fruitful in the study of problems of physics and chemistry associated with changes in intra- and inter-macromolecular interactions, structural and conformational transformations in macromolecules and complex polymer systems, relationships between the reactivity of macromolecules and their intramolecular mobility, etc.

The PL method is based on the study of polarization and other parameters of light emitted by luminescent groups (LM) covalently bonded to some parts of the macromolecules or polymer systems.

The PL method can be widely and systematically used for the study of polymers and polymer systems of various classes if it is based on specifically developed methods of covalent bonding of LM to the macromolecules of the polymers investigated. The structure and optical properties of luminescent groups, their amount and type of bonding to the polymer chain should meet certain requirements. Moreover, LM should not effect the investigated properties of macromolecules.

List of Abbreviations

PL	Polarized luminescence
LM	luminescent marker
LG	luminescent group
IMM	intramolecular mobility
ADM	9-anthryldiazomethane
PA	polyacid
PAA	polyacrylic acid
PAmA	polyamic acid
PGA	polyglutamic acid
PDME	polydimethoxyethylene
PMAA	polymethacrylic acid
PMA	polymethylacrylate
PMA-n	polyalkylmethacrylate-n
PMMA	polymethylmethacrylate
PPBE	polypropenylbenzyl ether
PS	polystyrene
$P(\alpha\text{-}CH_3S)$	poly-α-methylstyrene
PVBE	polyvinylbenzyl ether
PMMAA-n	methyl ⎱
PCMAA-n	cetyl ⟶ ester of poly-N-methacryloyl-ω-aminocarboxylic acids
PChMAA-n	cholesterol ⎰
PVpD	polyvinylpyrrolidone
VP	vinylpyrrolidone
VNH_2	vinylamine
CMC	carboxymethyl cellulose
PMMA-C	crosslinked polymethacrylic acid
PC	polymer-polymer complex
REE	rare earth element
DMF	dimethylformamide

1 Introduction

This review is divided into five sections. The first section deals with polarized luminescence as a physical phenomenon and with its phenomenological theory and the second
section with some methodological aspects taking into account the distinctive features
of the *PL* of macromolecules with LM.

Section 3 is concerned with the methods of the preparation of polymers with
covalently bonded LM that meet the requirements of the *PL* method and are located
in definite parts of macromolecules (in the main chain, at its ends, in side chains of
various structure or in certain structural fragments of branched chain and cross-
linked polymers).

Section 4 is devoted to some aspects of the experimental investigation of polymers in solution by *PL*. The results of the studies on the intramolecular mobility
(IMM) of various parts of the polymer chain and its relationship to the chemical
structure of macromolecules and the effect of hydrogen bonds and hydrophobic
interactions on the IMM are discussed. Special attention is devoted to the relationship between the IMM of polymers and the intramacromolecular structurization
caused by various factors (various specific intramolecular interactions or the presence
of surface-active substances and ions of rare earth elements). This section also deals
with the results of the investigation by the *PL* method of such complex and multicomponent polymer systems as block copolymers, cross-linked polymer systems and
intermacromolecular complexes.

The theoretical aspects of the micro-Brownian motion of polymer chains in
solution in connection with problems of the *PL* are dealt with in Sect. 5. They include the problems of the shape and width of relaxation spectra and the most probable relaxation times manifested in the motion of a given labelled chain element,
active in the *PL,* and the problems of the superposition of various types of motions
and the anisotropy of local relaxation properties etc.

The relationships of *PL* in the impulse regime are not considered. We will restrict
ourselves to the consideration of the main features of *PL* under stationary irradiation
since most experimental data obtained so far are based on this method.

Sections 2 and 4 were written by E. V. Anufrieva, Sects. 1 and 5 by Yu. Ya. Gotlib
and Sect. 3 by M. G. Krakoviak.

1.1 Polarized Luminescence

The complete description of luminescence as a complex electronic-optical phenomenon is possible only within the framework of quantum electrodynamics. These
problems are considered in well-known monographs by Vavilov[1], Levshin[2],
Feofilov[3], Stepanov[1, 5], Prinsgheim[6] and original papers by Levshin[7], Perrin[8] and
Weber[9, 10]. But as other optical phenomena occurring on the atomic or molecular
level (light absorption, emission or scattering, etc.), many aspects of luminescence
can be explained on the basis of classical (not quantum!) or semiclassical concepts
and models[5].

The classical oscillator model was found to be particularly useful in the description of the relationships of the polarization of the luminescence and its dependence on rotational Brownian motion and other factors for small molecules[11-14].

This fact permits the use of the methods and results of the oscillator model for the theory of PL for polymers containing luminescent groups.

This paper shows that of all properties of luminescence the polarization of luminescent light is the quantity the most sensitive to the micro-Brownian motion of macromolecules. Doubtless, many other parameters of luminescent light (including the shape of liminescent spectra, the quantum yield, the lifetime in the excited state, etc.) also depend on the molecular motion of luminescent groups and surrounding molecules of the medium. The motion of polar molecules of the medium (or molecules exhibiting some multipole moment) affects the local field near the emitter of the luminescent group thus changing the probabilities of absorption and secondary emission[15-19].

Polarized luminescence[1-8] is a radiation for which the amplitudes (or intensities) of vibrations of the light vector in two directions normal to each other and lying in a plane normal to the direction of the incident light are not identical. We will deal only with *PL* observed in the direction normal to the linearly polarized or natural incident light.

The investigation of the angular dependence of *PL* for different mutual orientations of the incident light and secondary emission (polarization diagrams) is a very important problem. The angular dependence provides information on the nature of elementary emitters[1-3, 20] (electric dipoles, magnetic dipoles, electric and magnetic quadrupoles and other multipoles) and the distribution of orientations of emitting and absorbing oscillators of luminescent groups in anisotropic media.

Levshin[2], Vavilov[1], Feofilov[3], Sevchenko and Sarzhevsky[21] and others have shown that, virtually in all the cases of *PL* investigated, experimental dependences correspond to the emission of a electric dipole oscillator.

The dipole nature of oscillators is also confirmed by other methods (for further details see Ref.[1]).

In general, a simple oscillator model of classical electrodynamics that ceases to be valid in some stage of the analysis of the spectra of atoms and simple molecules, can describe, at least formally, all the phenomena in the polarized liminescence of complex molecules in solution (Ref.[3], p. 121).

The degree of polarization is given by

$$P = (I_z - I_x)/(I_z + I_x) \tag{1.1.1}$$

if the incident light is directed along the *x* axis and the luminescence is observed along the *y* axis of the Cartesian coordinate system. Here, I_z and I_x are intensities of the components of luminescent light polarized in two directions normal to each other, I_z being polarized in the direction of the highest amplitudes of vibrations of the electric vector of the incident exciting light and I_x being polarized in the direction of the lowest amplitudes of vibrations. If the incident light is linearly polarized, I_z is measured along its polarization vector.

A simple relationship between the values of polarization for the illumination by polarized light P_p and for its excitation by natural light P has been derived by Vavilov and Levshin[1-5, 22, 23]

$$P = P_p/(2 - P_p)$$ (1.1.2)

Recently, another definition of the degree of polarization r for luminescence in a polarized exciting light differing from Eq. (1.1.2) has been proposed by Tao[24, 25] and Gordon[26]

$$r_p = (I_z - I_x)/(I_z + 2 I_x)$$ (1.1.3)

The measure of polarization determined by use of Eq. (1.1.3) is simply related to the dynamic properties of rotatory diffusion of luminescent groups. From Eqs. (1.1.3) and (1.1.1) we obtain

$$r_p = (2/3) [1/P_p - 1/3]^{-1}$$ (1.1.4)

If the exciting light is natural, the following value should be taken as the measure of polarization analogous to Eq. (1.1.3)

$$r = (I_z - I_x)/(2 I_z + I_x)$$ (1.1.5)

Then, from Eqs. (1.1.1) and (1.1.5) we obtain

$$r = (2/3)[1/P + 1/3]^{-1}$$ (1.1.6)

The use of the quantities I_p and r is based on physical rather than formal reasons, since these quantities are additive when several types of luminescent markers exist exhibiting luminescence in the same spectral range but characterized by different values of other parameters of PL (such as limiting polarization, times of rotational mobility and the duration of the afterglow).

If the luminescent light is to be polarized, it is necessary, first, that the emission of each elementary oscillator should be anisotropic in intensity, depending on the orientation of the emitter with respect to the direction of observation of luminescence, and the distribution of excited oscillators should be anisotropic with respect to the direction of observation.

Even for systems of emitters with isotropic distribution of orientations, the anisotropy of the excitation of emitters by polarized or natural light always occurs, being caused by the anisotropy of the incident light itself. The oscillations of the exciting electric vector always take place in the plane normal to the direction of the incident light and for linearly polarized light they occur in the direction of the polarization vector.

Anisotropy in the distribution of orientations of emitters always exists in macroscopic anisotropic media, crystals and media undergoing the effect of orienting external, mechanical, electrical or magnetic fields. Recently, the PL method has been successfully applied to the study of the properties of anisotropic media[27-32] (see also Refs.[27, 28] in connection with the study on the oriented-deformed state of polymers undergoing various mechanical loadings and strains (uni- and biaxial stretching) and polymers[33, 34] in an electric field and in flow[35-37]. The luminescent molecules placed in a strong electric field can be oriented owing to both a constant dipole moment and a dipole moment induced by the field.

The anisotropic distribution of emitters can be generated even in an isotropic medium by the action of polarized light[3, 38] if this medium exhibits photochemical activity, i.e. if photochemical reactions can take place in it at a rate depending on the angle between the orientation of a particle and the electrical vector of the incident light wave (Weigert effect). Photoselection takes place in the medium, i.e. the emitters of a certain orientation are selected by the light.

In this work, only the *PL* of emitters exhibiting initial isotropic distribution is considered. These emitters form a part of luminescent groups covalently bonded to macromolecules embedded in a viscous solvent.

1.2 Phenomenological Theory of Polarized Luminescence of Macromolecules with Luminescent Markers

This section only deals with some general concepts of molecular and phenomenological theory of *PL* and with the main theoretical relationships used in the following sections in which experimental data are reported.

The change in *PL* in the simplest case of the axially symmetrical Brownian motion of the emitting oscillator relative to the absorbing one is determined by Eqs. (1.1.1)–(1.1.6)

$$[1/P \pm 1/3] = [1/P_0 \pm 1/3] \left\{ (3/2 \, \tau_f) \int_0^\infty [\langle \cos^2\theta \rangle - 1/3] \exp(-t/\tau_f) dt \right\}^{-1} \quad (1.2.1)$$

where signs + and − refer to the excitation of luminescence by natural and polarized light, respectively; it is assumed that the observation is at an angle of 90° to the excitation direction.

The value of $\langle \cos^2 \theta(t) \rangle$ in Eq. (1.2.1) is the mean square cosine of the rotation angle θ of the emitting oscillator during time t, τ_f the average lifetime of the oscillator in the excited state with an exponential decay and P the value of polarization measured experimentally under steady-state irradiation. The value of τ_f represents the characteristic time scale with which the times of the micro-Brownian motion of the oscillator of a luminescent molecule or a luminescent group in the macromolecule are compared.

The value of P_0 is the limiting value of the polarization of luminescence observed when the Brownian motion is frozen, e.g. in very viscous media in which $(T/\eta) \to 0$ (where η is the viscosity of the solvent and T the absolute temperature[3-13]. This value is determined by the chemical structure of the luminescent macromolecule and the mutual arrangement of emitting and absorbing oscillators[1-3].

In the subsequent discussion, we will use the reduced measure of changes in polarization

$$Y = [1/P + 1/3]/[1/P_0 + 1/3] \qquad (1.2.2)$$

The relationships of PL as a function of kinetic properties of a luminescent particle are given by the time dependence $\langle \cos^2 \theta \rangle$.

The simplest case of the rotational Brownian motion of a spherical partice or a rigid dumbell in a viscous medium is given by[1-8]

$$\langle \cos^2 \theta \rangle = 1/3 + (2/3) \exp(-3 \, t/\tau_{rot}) \qquad (1.2.3)$$

where τ_{rot} is the time of rotatory diffusion

$$\tau_{rot} = (2 \, D)^{-1} \qquad (1.2.4)$$

and D the coefficient of rotatory diffusion. It should be noted that the relaxation of dipole polarization of the same particle with a dipole moment is represented by the value of $\langle \cos \theta \rangle$ and also characterized by the time τ_{rot}

$$\langle \cos \theta \rangle = \exp(-t/\tau_{rot}) \qquad (1.2.5)$$

i.e. the characteristic times for $\langle \cos \theta \rangle$ and $\langle \cos^2 \theta \rangle$ differ by a factor of 3.

Using the well-known Debye equation for τ_{rot}

$$\tau_{rot} = (3 \, \eta \, V/kT) \qquad (1.2.6)$$

where η is the viscosity of the liquid, V the particle volume and k the Boltzmann constant, we obtain the equation of Levshin-Perrin[7, 8]

$$Y = 1 + (\tau_f kT/\eta V) \qquad (1.2.7)$$

The analysis of experimental data on the micro-Brownian motion in polymer chains and the theory of relaxation phenomena in polymers (see Sect. 5) show that the Brownian motion of an oscillator in a luminescent marker covalently bonded to the chain obeys a more complex time law than Eq. (1.2.3). According to the theory of the relaxation processes, for a non-inertial physical system, the decay of $\langle \cos^2 \theta(t) \rangle$ will described by a spectrum of relaxation times τ_j[9-11, 39, 40] (or, more precisely, $\tau_j/3$ in order to retain the form of Eq. (1.2.3). In this case, we have

$$\langle \cos^2 \theta \rangle = \langle \cos^2 \theta \rangle_\infty + [1 - \langle \cos^2 \theta \rangle_\infty] \sum_j f_j \exp(-3 \, t/\tau_j) \qquad (1.2.8)$$

where f_j are relative weights of processes with times τ_j $(\sum_j f_j = 1)$, $\langle \cos^2 \theta \rangle_\infty$ is the mean square cosine of the rotation angle which is established for the equilibrium distribution of orientations (at $t \to \infty$) for the class of motions[40] under consideration. If this class includes motions leading to the isotropic distribution of orienta-tions of the oscillator in space at $t \to \infty$, it follows that $\langle \cos^2 \theta \rangle_\infty = 1/3$. In this case, isotropic rotatory diffusion of the particle occurs. However, dynamic models with anisotropic motion are also possible, e. g. the rotation of the oscillator about a fixed axis (or axes) when $\langle \cos^2 \theta \rangle$ is less than $1/3$.

If a completely isotropic distribution of oscillators (orientational randomisation) can be established we have

$$\langle \cos^2 \theta \rangle = 1/3 + 2/3) \; \Sigma_j \; f_j \; \exp(-3t/\tau_j) \tag{1.2.9}$$

By using Eq. (1.2.9) the most general relationship (1.2.10) between the reduced measure of the inverse value of polarization and the spectrum of relaxation times is derived.

$$Y = [1/P + 1/3]/[1/P_0 + 1/3] = \left\{ \Sigma_j f_j/(1 + 3 \; \tau_f/\tau_j) \right\}^{-1} \tag{1.2.10}$$

When $f_j = 0$ (if $j \neq 1$), $f_1 = 1$ and $\tau_1 = \tau$, Eq. (1.2.10) becomes an equation with one time of rotatory diffusion

$$Y = 1 + 3 \; \tau_f/\tau \tag{1.2.11},$$

and, correspondingly, if $\tau = \tau_{rot} = 3 \; \eta V/kT$ it becomes the Levshin-Perrin equation (Eq. (1.2.7.)).

Equation (1.2.10) has been derived by Weber[9] for a mixture of molecules (pro-teins) of various types with different rotatory diffusion times τ_j (or different vol-umes V_j) and the numerical contribution of a given sort f_j bearing identical lumines-cent markers with the duration of emission τ_f.

The theory of relaxation processes in macromolecules with internal rotation or torsional vibrations immersed in a viscous solvent with viscosity $\eta \sim 0.01$ P shows that, for local motions of a relaxation type with characteristic times $\tau > 10^{-9}$ s, the so-called condition of high friction is satisfied in a vast majority of cases. Relaxation times are given by the equation

$$1/\tau_j = a_j(T/\eta) \exp[-U_j/kT], \tag{1.2.12}$$

where U_j are barriers to the internal motion in the chain or in side groups, or by the equation[40, 41-46]

$$1/\tau_j = kT/3 \; \eta \; (V_j)_{eff} \tag{1.2.13}$$

where $(V)_{eff}$ is the effective "volume" of the particle. In this case, when the analysis of the experimental data is based on the generalized Levshin-Perrin-equation (1.2.10), it is also useful to introduce the dependences $(1/P)$ or $Y = Y(T/\eta)$.

When Eqs. (1.2.12) and (1.2.13) are obeyed, the dependences $Y(T/\eta)$ have the shape of curves convex toward the positive ordinate. If U_j in Eq. (1.2.12) is equal to

0 (or if the effective volumes are independent of temperature), then Y is a unique function of ratio T/η only and isotherms $Y(T/\eta)$ for different temperatures coincide (if the solvent viscosity is varied). In the opposite case (at $U_j \neq 0$), the dependences $Y(T/\eta)$ obtained when the temperature (for a given solvent) is varied will differ from those obtained when η is varied and T = const and different isotherms will differ from each other.

In a general case, the reconstruction of the form of the time dependence of function $(3/2) [\cos^2 \theta(t) - 1/3]$, i.e. of the mean square cosine of the rotation angle of the oscillator, and a direct determination of the shape of the relaxation time spectrum (all f_j and τ_j) from the experimental curve $Y(T/\eta)$ are complicated mathematical problems (see Sect. 5). They may be solved rigorously (under the condition that $\tau_j \sim a_j(\eta/T)$) only if the dependence $Y(T/\eta)$ is known over the entire range of changes in T/η (from $T/\eta = 0$ to $T/\eta \to \infty$). Doubtless, this is impossible in practice.

However, it has been established by Weber[9] (see also Ref.[39]) that some characteristic averaged parameters of the relaxation spectrum can be obtained more or less simply from the analysis of the behavior of $Y(T/\eta)$ over a limited range of T/η values (in particular, at $T/\eta \to 0$) or, if certain assumptions are made, concerning the asymptotic behavior of $Y(T/\eta)$ at $T/\eta \to \infty$ (by the extrapolation of experimental data in a finite range of T/η).

For very viscous solvents, i.e. at low values of T/η at which even for short times of the relaxation spectrum, τ_f/τ_j is less than 1, it is possible to expand $1/P$ (or Y) in power series of T/η (or of τ_f/τ_j) (cf.[39]) and to determine the initial slope of curve $Y(T/\eta)$

$$Y \cong 1 + 3 \tau_f \sum_j (f_j/\tau_j) - 9 \tau_f^2 [\langle 1/\tau^2 \rangle - \langle 1/\tau \rangle^2] + \ldots \tag{1.2.14}$$

The value of the reciprocal average relaxation time is given by the expression

$$(1/\tau_{-1}) = \langle 1/\tau \rangle = \sum_j f_j/\tau_j = (T/3 \tau_f \eta) [\partial Y/\partial (T/\eta)]_{\frac{T}{\eta} \to 0} \tag{1.2.15}$$

On the other hand, by using Eq. (1.2.10) it is possible to obtain the expression of the asymptotic dependence $1/P$ for low viscosity solvents at $T/\eta \to \infty$, i.e. for so high values of $\tau_f T/\eta$ at which, even for the highest values of τ_j, the condition $\tau_f/\tau_j > 1$ is obeyed. In this case, it is possible to obtain an asymptotic expansion for Y

$$Y \cong \langle \tau^2 \rangle / \langle \tau \rangle^2 + 3 \tau_f/\langle \tau \rangle - [\langle \tau \rangle \langle \tau^3 \rangle - \langle \tau^2 \rangle^2]/3 \tau_f \langle \tau \rangle^3 + \ldots \tag{1.2.16}$$

which contains the number-average time of the relaxation spectrum

$$\tau_n = \langle \tau \rangle = \sum_j f_j \tau_j \tag{1.2.17}$$

and the mean square time $\sqrt{\langle \tau^2 \rangle}$

$$\langle \tau^2 \rangle = \sum_j f_j \tau_j^2 \tag{1.2.18}$$

Since the dependence $\tau(\eta/T)$ is linear, the asymptotic behavior of the value of $Y(T/\eta)$ at $T/\eta \to \infty$ is given by

$$Y\bigg|_{\frac{T}{\eta} \to \infty} = Y_0' + C \cdot T/\eta \qquad (1.2.19)$$

where the slope of the linear dependence determines the average relaxation time (at a given value of T/η):

$$C \cdot T/\eta = 3 \tau_f/\tau_n = 3 \tau_f/\langle \tau \rangle \qquad (1.2.20)$$

and the intersection of the asymptote with the ordinate gives the intercept Y_0', equal to the ratio of $\langle \tau^2 \rangle$ to $\langle \tau \rangle^2$

$$Y_0' = \langle \tau^2 \rangle / \langle \tau \rangle^2 \qquad (1.2.21)$$

The value of the intercept is also a measure of the width of the distribution of relaxation times, which is the most sensitive to the distribution of long times. It is convenient to derive the weight-average relaxation time

$$\tau_w = \langle \tau^2 \rangle / \tau_n = \langle \tau^2 \rangle / \langle \tau \rangle \qquad (1.2.22)$$

Hence,

$$Y_0' = \langle \tau_w \rangle / \tau_n \qquad (1.2.23)$$

It can be seen that this is the usual measure of the distribution width, e.g. in the theory of molecular weight distributions.

Generally speaking, a linear asymptote at $T/\eta \to \infty$ occurs only if relaxation time spectra are not very broad or are limited. In this case the values of $\langle \tau^2 \rangle$ and $\langle \tau \rangle$ do not become infinity (see Sect. 5.3). In principle, such theoretical distributions of τ_j for which the asymptotic behavior at $T/\eta \to \infty$ is not linear are also possible.

When a linear asymptote really exists, it is convenient to compare reciprocal polarization with the value of $1/P_0' + 1/3$, the intercept formed on the ordinate by the asymptote when it is extrapolated to $T/\eta \to \infty$ rather than with the true limiting reciprocal polarization $1/P_0 + 1/3$. The reduced reciprocal polarization can be defined as

$$Y_0' = (Y - Y_0')/Y_0' = [1/P - 1/P_0']/[1/P_0' + 1/3] \qquad (1.2.24)$$

Hence, according to Eqs. (1.2.19), (1.2.23) at $T/\eta \to \infty$ we obtain

$$Y' \cong 3 \tau_f/\tau_w \qquad (1.2.25)$$

Consequently, the slope of the asymptotic linear part of the curve in the system of coordinates $Y'(T/\eta)$ will determine the value of τ_w rather than τ_n.

In a general case, when the asymptote is approached, the following conditions should be satisfied: $\tau_f/\tau_n > 1$ and $\tau_f/\tau_w > 1$. However, in practice the linear dependence $Y(T/\eta)$ at high values of (T/η) over a finite range of T/η can also be observed at the values of τ_n and τ_w comparable to τ_f. In this case, the characteristic times determined from the slope of $Y(T/\eta)$ or $Y'(T/\eta)$ (by using Eq. (1.2.25)) are not true average values (of the type of τ_n or τ_w) over the entire range of the relaxation spectrum, but have a different physical meaning.

Thus (compare Sect. 5.5), a situation can be envisaged, in which the relaxation spectrum includes both such rapid high-frequency motions with $\tau_{h.f.}$ that $\tau_f/\tau_{h.f.} > 1$ over the entire experimental range of T/η (except its lowest values) and relatively slower motions with $\tau'_j \gg \tau_{h.f.}$ for which, however, the condition $(\tau_f/\tau'_j) > 1$ is also fulfilled over the given range of T/η.

In this case, the dependence $1/P$ (or Y vs T/η) over the range attained experimentally is described by a straight line (a "pseudo-asymptote") and its slope is determined mainly by times τ'_j.

This pseudo-asymptote intersects the ordinate at point P'_0 that corresponds to the contribution to polarization provided by rapid processes occurring at very low T/η. In accordance with Eq. (1.2.27), the value of $(\tau_w)_{eff}$ formally determined from the linear dependence considered above has the sense of τ_{-1} averaged only over the slower processes τ' with renormalized weights related only to these processes

$$(1/\tau_w)_{eff} = 1/\tau'_{-1} = \sum_j \varphi'_j / \tau'_j \qquad \sum \varphi'_j = 1 \qquad (1.2.26)$$

and the value of

$$Y'_0 \cong [1/P'_0 + 1/3]/[1/P_0 + 1/3] = (1 - f_{h.f.})^{-1} \qquad (1.2.27)$$

depends on the contribution of high-frequency processes $f_{h.f.}$ to the initial relaxation spectrum. It is clear that now $(\tau_w)_{eff}$ can be lower than τ_f.

Below we use the actual values of $(\tau_w)_{eff}$ determined from the linear part of the dependence $P^{-1}(T/\eta)$ which, in each case, depending on the ratio of τ_w to τ_f, represent different effective average values of τ_j.

2 Polarized Luminescence (PL) of Macromolecules with Conjugated Luminescent Markers

The theoretical analysis of the relationship between small-scale relaxation processes in polymers (PL reveals just these processes) and the intra- or intermolecular interactions[47] shows that these processes are sensitive to various changes in the number (probability of formation) of intra- or interchain contacts in polymer solutions. The experimental data reveal that the intramolecular mobility (IMM) of the polymer in solution is highly susceptible to various conformational transformations of macromolecules and to changes in the intermolecular interactions in complex multicomponent polymer systems[40, 48]. Therefore, the *PL* method is very efficient for solving many problems of polymer chemistry and physics. The investigation of polymer solutions becomes important for the understanding of not only the structure and properties of single polymer chains but also of polymers in a condensed state.

The high effectiveness of the *PL* method for solving various problems of polymer chemistry and physics is mostly due to the features that distinguish this method from other methods applied to the investigation of the structure and properties of polymers in multicomponent polymer systems in solution.

These features are as follows:

1. The possibility of carrying out investigations in very dilute solutions at a polymer content in solution as low as 0.001%. This means that the effect of interchain contacts superimposed on intramacromolecular interactions is greatly weakened or completely absent. This weakening is particularly important when the polymer-polymer interactions sharply increase, e.g. by the action of the precipitant;

2. The *PL* method permits to carry out investigations in any solvents: polar and non-polar solvents, in water, in the presence of charged particles and at any temperatures;

3. The *PL* method permits the investigation of polymers of any chemical structure: linear or branched soluble polymers and even cross-linked insoluble polymers[49]. The only exceptions are polymers extinguishing luminescence. However, this property permits the development of new approaches based on this property[50]. For cross-linked polymers it was found sufficient to disintegrate the polymer into particles of a uniform size not exceeding 0.5μ. This approach is widely used in the investigation of polyelectrolytic networks[49, 51].

4. Finally, the *PL* method is useful for a study of individual components in multi-component polymer systems under the same experimental conditions. This problem can be solved by using conjugated luminescent samples (or markers); it is possible to distinguish the component investigated from other components by combining it chemically with a liminescent sample. Therefore, the *PL* method gives much information on the structure and properties of block copolymers (such as three-block copolymers), on the formation and functioning of the polymer-polymer complexes and on the competing interactions in multicomponent systems.

The use of labels makes it possible not only to distinguish a single component in a multicomponent polymer system, but also to distinguish a single part of the macromolecule: the end of the main chain, the end of the side chain, the inner part of the main chain. Separate investigations of the mobility of the main chain and side chains are particularly desirable for polymers with a complex comb-like structure.

Doubtless, to utilize completely all the advantages of the *PL* method, the team work of physicists and chemists is necessary as well as the efforts in the field of synthetic chemistry aimed at developing such methods of the synthesis of polymers with markers that can be used for polymers of any chemical structure. At present, this problem has been completely solved for antbracene-containing markers by Krakoviak (Table 1). Section 3 deals with the methods of synthesis of labeled polymers.

Table 1. Structure and designation of luminescent markers (LM)

It is desirable to introduce labels of the same structure into all investigated polymers. This means that for all these polymers the contribution of the marker to the relaxation spectrum of the polymer is the same. The experiment shows that under these conditions it possible to observe even the smallest differences in the IMM of polymers and to draw conclusions concerning slight changes in their structure and properties. The use of markers with anthracene structure also offers certain advantages for the investigation of the IMM of polymers. First, in organic solvents and in water the anthracene ring (in contrast to aromatic rings with chromophore groups) does not form any specific bond with the groups of the polymer apart from hydrophobic interactions in water that can easily be taken into account. Second, owing to the presence of the bulky anthracene ring, low-frequency motions in the side unit containing luminescent groups are also relatively hindered so that the IMM of the labelled polymer is mainly determined by the IMM of the main chain. The markers with the anthracene structure (for an anthracene ring bonded to the chain in the position 9) are remarkable in that the high-frequency motion of the luminescent group (LG) itself about the bond adjoining the anthracene ring is not shown in PL because the direction of the dipole moment of the transition coincides with the direction of this bond[52–54].

The results of the theoretical analysis in Sect. 1.2 show that in the study of relaxation processes in solutions of polymers containing luminescent markers (LM) much information is provided by the dependence of the reciprocal value of the polarization of luminescent light of the polymer solution, $1/p$, on the viscosity of the solvent η, $1/p = f(T/\eta)$, measured at constant temperature T. Theoretical analysis of Eq. (1.2.10) shows that for macromolecules in solution in which all the relaxation processes occur at times $\tau_j > \tau_f$ (τ_f is the lifetime of luminescence) the following equation is valid for the average time τ_{-1} (see Sect. 1.2) (Eq. 2.1)[39, 40].

$$1/\tau_{-1} = \sum_j f_j/\tau_j \qquad (2.1)$$

This is the range of the initial slope of the isothermal dependence $1/p(T/\eta)$. To find τ_{-1} it is sufficient to evaluate the slope of the tangent to curve $1/p = f(T/\eta)$ at $T/\eta \to 0$. When the viscosity of the solvent η decreases, τ_j also decreases since $\tau \sim \eta$[55,56]. At low values of η or at high values of T/η, only the part τ_j' is greater than τ_f and on the curve $1/P = f(T/\eta)$ the initial linear part extrapolated at $T/\eta \to 0$ to $1/P_0$ is replaced by a part with a lower slope extrapolated at $T/\eta \to 0$ to $1/P_0'$. This part of the dependence $1/P = f(T/\eta)$ permits to determine τ_w characterizing the intramolecular mobility of the polymer chain of high molecular weight $M > 10^4$ (for anthracene-containing markers)[40, 48] by Eq. (2.2)

$$\tau_w = \frac{(1/P_0' + 1/3)\, 3\, \tau_f}{1/P - 1/P_0'} \qquad (2.2)$$

Subscript "w" means that the times were determined in the range of the linear part of the $1/P(T/\eta)$ dependence at high $T/\eta > (1 - 2) \times 10^{+4}$ K/P^{+1} $1/P_0'$ is the parameter reflecting the contribution of the amplitude of high-frequency twisting vibrations in the unit with LM to the change in PL of the solution of the polymer with LG.

In previous publications[40, 48, 57], the relationship between $1/P_0'$ and the chemical structure of the bridge bonding the anthracene ring to the polymer (Table 2) on the one hand, and the conformational transformations of macromolecules on the other hand (Tables 3 and 4) have been studied. It is shown that the value of $1/P_0'$, does not depend on the nature of the solvent in the absence of conformational transitions the polymer and markedly changes with the internal structure of the polymer (Tables 3 and 4) and the chemical structure of the bridge bonding the anthracene ring to the polymer (Table 2). The theoretical analysis of this relationship permits the evaluation of the amplitude of high-frequency twisting vibrations of the anthracene ring for different types of bonding of this ring to the polymer[58]. The value of $1/P_0'$ increases with temperature (Fig. 1).

In order to obtain the dependence $1/P = f(T/\eta)$ at T = const. by using various solvent mixtures, the following requirements should be fulfilled: 1) the corresponding changes in τ_f and 2) the conformational transformations of macromolecules occurring upon variation of the composition of the solvent mixture[56] should be taken

Fig. 1. Plot of $\dfrac{1}{P_0'}$ vs. temperature for solutions of polymers containing as LM$_1$ marker 9-anthryl methyloxy-carbonyl group. ▲ poly(methyl methacrylate) in methyl acetate-triacetin; ○ poly(hexadecyl methacrylate) in toluene-cyclohexanol

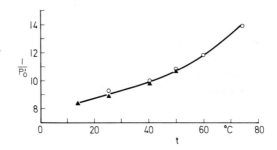

Table 2. Relaxation times for conjugated polystyrene with various luminescent markers in toluene at 25 °C and values of $\frac{1}{P_0'}$ and $\frac{1}{P_0''}$ (Symbols for types of LM as in Table 1)

Type of LM	$\frac{1}{P_0'}$	$\frac{1}{P_0''}$	τ_W (ns)
LM$_8$	12	40	5.1
LM$_4$	20	36	5.0

5.2

16

9

LM$_1$

8.9

13

7

LM$_5$

4.1

27

12

LM$_{10}$

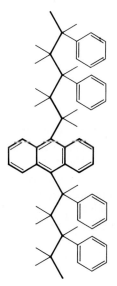

Table 3. Values of $1/P_0'$ for various polymers with LM_1 and LM_5 markers at 25 °C. Luminescence excitation by unpolarized light (α is the degree of ionization.) $\lambda_{exc.}$ = 365 nm

Polymer	Solvent	$1/P_0'$ for LM_1	$1/P_0'$ for LM_5
PMMA	Methyl acetate	9.0	7.0
PMAA[a]	Methanol	9.0	7.0
PAA[b]	Water	9.0	–
PMAA, $\alpha = 0$	0.002 N/HCl-water	6.5	5.7

[a, b] PMAA (PAA) – poly(methacrylic or acrylic acid)

Table 4. Value of $\dfrac{1}{P_0'}$ vs. pH for aqueous solutions of poly(glutamic acid) at 25 °C. LM_3 marker
with the structure

$$\text{—CH}_2\text{—CH}_2\text{—}\overset{\overset{\displaystyle O}{\|}}{C}\text{—O—CH}_2\text{—}$$

Luminescence excitation by unpolarized light, $\lambda_{exc.}$ = 365 nm

pH	$\dfrac{1}{P_0'}$
6.1	16
6.0	13.9
5.4	13.9
5.2	10.9
4.6	10.6
4.3	10.5

into account. Requirement 1) is met by direct measurements of τ_f with a phase fluorometer of high precision (2–5% of the measured value) for all mixed solvents. Table 5
shows the dependences of τ_f on the composition and nature of solvent, the type of
bonding of the marker to the polymer chain and the transformations of the intramacromolecular structure[48, 57]. The structure of LM used in Refs.[48, 57]and
the designations employed in these papers are listed in Table 1. Requirement 2 is
fulfilled by taking into account such conformational changes in macromolecules
occurring by the action of mixtures of solvents of different compositions that has
become apparent at least in the changes of intrinsic viscosity $[\eta]$ of the polymer
solution. It has been shown by PL[56] and dielectric relaxation of dipole polarization[59]
that for PMMA and for all the polymers studied the ratio $\tau_{IMM} \sim 1/[\eta]$ is valid in
solvents of different thermodynamic strengths: the IMM of the polymer varies pro-

Table 5. Values of τ_f (ns) for polymers with various LM. (Symbols for types of LM as in Table 1)

Type of LM	τ_f(ns)	Polymer	Solvent
LM_1	4.3	PMMA	Methyl acetate
LM_1	4.8	PMAA[a]	Methanol-water (60:40)
LM_1	9.0	PMAA	Water
LM_2	8.9	PMMA	Methyl acetate
LM_4	3.4	PDME[b]	Methyl acetate
LM_5	7.2	PMMA	Methyl acetate
LM_6	7.6		
LM_7	5.2	PMAA	Water
LM_8	6.9	PS[c]	Toluene
LM_9	3.8	PMMA	Methyl acetate
LM_{10}	5.8	PMMA	Methyl acetate
LM_{11}	8.6	PEG[d]	Water

[a] PMMA poly(methacrylic acid)
[b] PDME poly(dimethoxyethylene)
[c] PS polystyrene
[d] PEG poly(ethylene glycol)

Table 6. Influence of solvent on relaxation times τ_w^{red} of PMMA with 9-anthrylmethyloxycarbonyl markers-LM_1 at 25 °C. $\eta_{red} = 0.38$ cP

Solvent	$[\eta]$ (dl/g)	τ_w^{red} (ns)
Dichloroethane	1.16	2.2
Methyl acetate	0.68	3.9
Toluene	0.61	4.3
Ethylene acetate	0.55	4.8
Propyl acetate	0.47	6.3
Methyl ethyl ketone/ 2-propanol (1:1)	0.38	—
Butyl acetate	0.32	7.8

Superscript "red" means that the values of $\tau_w \sim \eta$ are reduced to the same value of the viscosity of the solvent $\eta_{red} = 0.38$ cP (in order to take into account the influence of the viscosity of the solvent on τ_w)

portionally to the change in $[\eta]$ (Table 6). This is understandable because with increasing $[\eta]$ the amount of intramolecular collisions affecting the small-scale relaxation processes in chains decreases.

Times τ_w for phenyl-containing polymer molecules with anthracene-containing LM are determined by Eq. (2.3)

$$\tau_w = \frac{(1/P_0' + 1/3)\, 3\, \tau_f}{1/P - 1/P_0''} \tag{2.3}$$

$(1/P_0''$ is the parameter characterizing the conformational depolarization of luminescence[60], Table 2).

3 Methods for the Preparation of Polymers with Covalently Attached Luminescent Markers of the Anthracene Structure

The investigation of the intramolecular mobility (IMM) of macromolecules by the method of polarized luminescence (*PL*) is based on the study of the characteristics of luminescent light (polarization of luminescence and lifetime in the excited state) of luminescent groups bonded to the investigated macromolecules.

In general, the most comprehensive and accurate information concerning various relaxation processes typical of a given polymer and determining its characteristic IMM can be obtained by studying polymers that contain luminescent groups (luminescent markers, LM or "reporters") purposely covalently bonded to definite parts of the macromolecules. Usually, the LM content in the polymer does not exceed 0.1 mol% (i.e., 1 LM per 1000 units of the polymer of the main structure). An increase in the LM content can obscure the properties of the macromolecule.

It was necessary to carry out systematic investigations of the polarization of the luminescence of groups bonded to macromolecules and of its relationship to the peculiarities of the IMM of polymers of various structure. Hence, it was necessary to develop various methods for the synthesis of labeled polymers differing not only in the structure but also in the content and arrangement of LM in macromolecules (in side groups, in the main chain or at its end). Studies on the polarization of luminescence of LM with different locations in macromolecules permit to determine the dynamic characteristics of various parts of the polymer chain.

In the preparation of a system of labeled polymers, it is advisable to use luminescent groups exhibiting the same chemical structure as that of the main type of markers. If necessary, the study of polymers containing the main type of LM may be supplemented by an investigation of the same polymers with a LM of another chemical structure. The authors used an anthracene group as the main type of LM for the following reasons.

Anthracene groups attached to polymer chains as LM exhibit a combination of optical properties required for the study of the relaxation properties of polymers by the *PL* method. They are characterized by relatively high values of the limiting polarization and the quantum yield of luminescence. Absorption and luminescence bands of anthracene groups occur in the spectral range convenient for experiments. It is also important that the optical properties of LM with the anthracene structure do not vary greatly over a wide range of changes in the properties of the medium (pH of solution, solvent nature and structure of polymer units surrounding LM).

On the other hand, the structure of anthracene and its derivatives determines the peculiarities of their chemical behavior and the ability to participate in different chemical processes.

The anthracene group readily participates in various homo- and heterolytic substitution or addition reactions[61]. The meso-positions (i.e. positions 9 and 10) of the

anthracene system are the most reactive, but reactions involving other positions of
the anthracene molecule are also possible. The presence of an anthryl moiety in
compounds of the $A-CH_2-X$ type increases the chemical reactivity of the

$-CH_2-X$ functional group. The chemical properties of anthracene and its derivatives make it possible to synthesize anthracene-containing reagents and monomers
of various classes and to vary widely the methods of the preparation of labeled
polymers. However, these properties require great care in the preparation of anthracene-containing polymers, since undesirable side reactions involving anthryl groups
occur quite readily; they can lead to inhibition of the polymerization and to branching or cross-linking as well as to changes in the structure and, hence, in the photophysical properties of LM.

Various methods of synthesis permit the preparation of polymers with different
structures and different contents and the required location of LM in macromolecules. For polymers with LM in side groups, it is possible to vary the structure of
the bridge between the luminescent group and the polymer chain and the position
of its bonding relative to the anthracene ring. This allows to change not only the
flexibility of the bridge but also the angle formed by the rotational axis of the
emitting oscillator and the direction of the oscillator itself.

LM can be introduced into macromolecules in all the stages of their formation
(initiation, propagation or termination) as well as in the reactions involving functional groups of preformed or natural polymers. The choice of the method for the
synthesis of a labeled polymer depends on its chemical structure and the required
arrangement of LM in the polymer chain.

3.1 Methods for the Preparation of Polymers with Luminescent Markers of the Anthracene Structure at Their Chain Ends

Polymers with LM at their chain ends can be prepared either by the synthesis of
macromolecules (in initiation or termination reactions) or by reactions involving
terminal functional groups of macromolecules.

The polymerization of methyl methacrylate initiated by the products of the
thermal degradation of 9-anthryldiazomethane can serve as an example of the synthesis of a polymer with terminal LM using an initiator-containing luminescent
group[62, 63].

The formation of polymers with terminal LM during chain termination in free-
radical polymerization is based on the ability of anthracene and some of its derivatives to participate in homolytical reactions[64, 65]. It was established that anthracene-
containing compounds interact with macroradicals which are generated in free-radical

polymerization of some monomers (vinyl acetate, methyl acrylate, styrene etc.) forming units exhibiting 9,10-dihydroanthracene structure I[66] (addition or copolymerization reaction, Scheme 1) or terminal anthryl groups II[67-69], i.e. LM (substitution reaction).

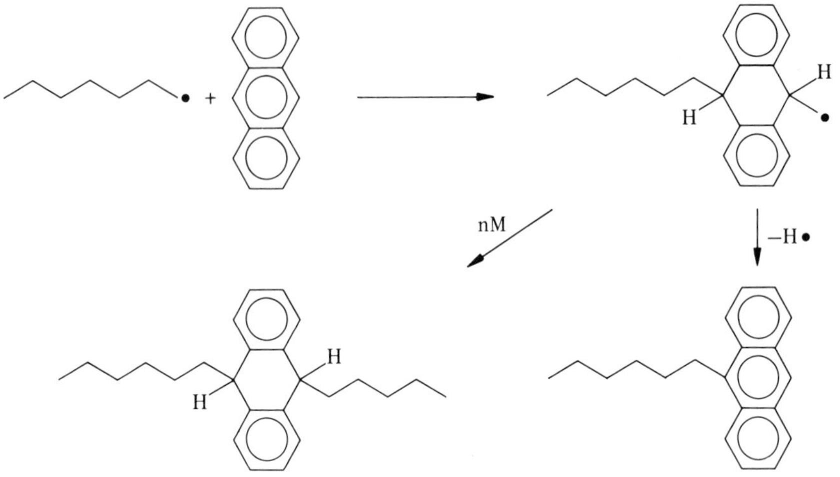

Scheme 1   I   II

The contribution of the homolytical substitution to the total sum of reactions occurring in free-radical polymerization in the presence of anthracene of its derivatives depends on the nature of the monomer and the anthracene-containing compound and on the polymerization conditions. It has been shown that the macroradicals of the methacrylic esters do not interact with anthracene under usual conditions[63, 66, 67]. On the other hand, in the polymerization of styrene, substitution with the formation of terminal LM of type II (Scheme 1) proceeds only if one or both meso-positions of the anthracene ring are free[67, 69].

Polymers containing LM at the chain ends can be obtained by the reaction of anthracene-containing reagents with functional end groups of macromolecules such as carboxy groups[70] or "living" ends of macromolecules generated in anionic polymerization[71].

3.2 Methods for the Preparation of Polymers with Luminescent Markers of the Anthracene Structure in Side groups of Macromolecules

Three types of methods are applied to obtain macromolecules containing LM in side groups:

3.2.1 Methods based on the copolymerization of the main monomer with the "label" amount of the monomer bearing the liminescent group.

3.2.2 Methods based on reactions involving macromolecules and reagents containing anthryl groups.

3.2.3 Combined methods.

3.2.1 Methods Based on Copolymerization

The possibility of obtaining a wide range of polymers with anthracene LM in side groups differing both in the structure of the bridge between the luminescent group and the main chain and in position of its bonding is determined by the synthesis of monomers bearing anthryl groups Monomers of the vinyl-aromatic type: 1-vinyl-anthracene, 2-vinylanthracene and 9-vinylanthracene are anthracene-containing monomers with the simplest structure[72, 73]. It is possible to use 1- and 2-vinylanthracenes for the introduction of LM into polymers formed by free-radical or ionic polymerization[73, 74]. In contrast to 1- and 2-vinylanthracenes, the homo- and copolymerization of 9-vinylanthracene proceeding by various mechanism is accompanied by isomerization and formation of 9-methylene-9,10-dihydroanthracene structure (Sect. 3.3, Scheme 2)[75–77]. These processes are caused by electronic and steric factors due to the location of the vinyl group at C-9 of the anthracene ring. Electronic and steric interactions between the vinyl group and the anthracene system are different in monomers derived from the 9-anthrylstyryl group[79, 82] in which the non-coplanar arrangement of phenyl rings with respect to the anthracene ring[78] distorts the conjugation between the vinyl and the anthracene groups. Hence, copolymerization with these monomers proceeds readily by a free-radical or ionic mechanism and is not accompanied by isomerization[79–81].

Anthracene-containing monomers of the methacrylic series (e.g. anthrylmethyl methacrylates)[63, 83–85] can also be used to obtain polymers with LM by copolymerization. The monomers of this type are successfully used for the introduction of LM (Table 1; LM_1 and LM_2) into side groups not only of various methacrylate polymers (e.g. from methyl to cetyl- and cholesteryl methacrylates)[63, 86–88], but also of other polymer types. In the latter case, it is advisable to use monomers in which the position 10 is substituted (e.g. by a methyl group). In this case, the probability of side homolytical reactions with the participation of anthracen groups consinderably decreases[68].

9-Anthrylmethyl vinyl ether obtained by trans-vinylation[89] is an anthracene-containing monomer which copolymerizes by a cationic mechanism. It has been successfully applied to the synthesis of polymers of mono- and dialkoxyethylenes and propenyl alkyl ethers bearing LM in side groups (Table 1; LM_4)[89].

Polymers with LM in side groups can also be obtained by polymerization of other monomers. Thus, poly(ethylene oxide) containing LM (Table 1; LM_6) is formed by copolymerization of ethylene oxide and 9-anthryaldehyde in the presence of a modified organoaluminum catalyst[90].

The preparation of "labeled" polyalkylidenes is another example[63]. Polyalkylidenes are hydrocarbon polymers of the $\{CH(R)\}_n$ type formed by polymerization or copolymerization of diazoalkanes $RCHN_2$; they are of interest as model polymers[91, 92]. Polyalkylidenes of various structures containing LM(Table 1, LM_6) can be obtained by copolymerization of diazomethane (or its mixture with other diazoalkanes) and 9-anthryldiazomethane[63].

3.2.2 Methods Based on Reactions with the Participation of Macromolecules

The methods based on reactions with the participation of macromolecules allow luminescent markers to be bonded to polymers and copolymers of the required composition, stereoregularity, molecular weight and molecular weight distribution, to samples of industrial polymers and to natural polymers.

In each case, the choice of the type of reaction between the macromolecules of the polymer and anthracene-containing reagents is determined by many factors. The most important include:

1. The existence, nature and reactivity of functional groups in the polymer molecules.
2. The existence of only one type of bonds between the macromolecule and the luminescent group. The formation of comparable amounts of different bonds affecting the optical and dynamic properties of LM in different ways can make difficult or prevent a quantitative interpretation of data obtained by the *PL* method.
3. The necessity of maintaining the functional and structural properties of the polymer.
4. The stability of the newly formed bond between the luminescent groups and the polymer chain under conditions of the subsequent experiment.

At present, methods have been developed permitting the bonding of LM with the anthracene structure to different polymer molecules. The reaction of 9-anthryl-diazomethane (ADM) or its derivatives with the COOH groups of macromolecules is the most convenient method for the attachment of LM (Table 1; LM_1, LM_2, LM_3 and LM_9) to carboxy-containing polymers[63, 93].

Just as most other diazoalkanes, ADM can readily react with carboxylic acids under mild conditions (in the absence of the catalysts, at or below room temperature and at low concentrations of the reacting compounds) forming the corresponding esters. The reactions between carboxycontaining polymers and ADM can be separated into three groups: 1) reactions carried out under homogeneous conditions, 2) reactions proceeding at the liquid-liquid interface and 3) reactions at the liquid-solid interface.

When 9-anthryldiazomethane and the polymer have a common solvent, the homogeneous method is the most convenient one, since it ensures the highest yield and leads to the most homogeneous distribution of LM in the polymer (in the case when the contents of carboxylic groups in macromolecules greatly exceeds the required amount of LM)[63, 94, 95].

The nature of the solvent plays an important role in the reactions between carboxy-containing compounds and diazoalkanes. Protic solvents of the R–OH type can react with diazoalkanes in the presence of carboxylic acids to form the corresponding ethers[96]. Hence, in reactions with polymers dissolved in alcohols or alcohol-containing mixtures (poly(acrylic and methacrylic acids) etc.) the yield of the main reaction can decrease to 30–50% (with respect to ADM)[63]. Consequently, when polymers soluble in aprotic solvents undergo anthrylmethylation, the reaction should be carried out in such solvents as benzene or toluene.

For water-soluble natural or synthetic polymers, one of the heterophase methods can be used[63]. In this case, a considerable portion of 9-anthryldiazomethane is consumed by side reactions and this loss must be considered when choosing the amount of the reagent.

The bonding of LM by using ADM to cross-linked carboxy-containing polymers swelling in organic solvents is an intermediate case[49].

It should be noted that anthrylmethylcarboxylate bonds formed in the reaction of ADM with the COOH groups are relatively stable in aqueous solutions of labeled polyacids or polypeptides. However, under the corresponding conditions ester bonds of this type can be readily split[97].

These methods of the reaction between 9-anthryldiazomethane and carboxylic groups of macromolecules permit to bond LM to side groups of synthetic and natural polymers and copolymers of acrylic and vinylbenzoic acids[98, 99], hydrolyzed copolymers of maleic anhydride[48], carboxymethylcellulose[48], comb-like alkyl acrylates[100], polyglutamic acid and other synthetic polypeptides[101] as well as proteins, such as invertase or ribonuclease[48].

9-Anthryldiazomethane is also used in another "carbene" method of LM bonding to polymers including polymers whose macromolecules do not contain reactive functional groups. It is known that carbenes are formed in the thermal or photolytic decomposition of diazoalkanes. They are highly reactive compounds of the general formula $R^1-\ddot{C}-R^2$ [92, 102], capable of undergoing various chemical reactions including insertion into C–H bonds.

Investigations have shown that 9-anthrylcarbene[103], generated by thermal or photolytic decomposition of 9-anthryldiazomethane in a polymer solution, can be inserted into the C–H bonds of the macromolecules[63]. As a result, anthracene groups are covalently bonded to the macromolecules (Table 1; LM_{11}).

Usually, the amount of anthracene groups bonded to the macromolecules during the carbene reaction is small. Under the conditions employed, it does not exceed 10–15% of the initial amount of ADM and greatly depends on the polymer structure, the nature of the solvent, and the polymer concentration[63].

The reaction of 9-anthrylcarbene with the polymers can affect the C–H bonds of various groupings of the macromolecules. Hence, if a predominantly identical structure of the bridge between the anthracene group and the polymer chain is required, it is advisable to use the carbene method for bonding the LM to polymers with one type of C–H bonds (polyethylene, poly(ethylene oxide) etc.).

The method based on the reaction between 9-chloromethylanthracene and hydroxy groups of the macromolecules can be used to bond LM (Table 1; LM_4) to polymers and copolymers of vinyl alcohol, cellulose derivatives, etc.

The same reagent containing a very reactive chlorine atom can be successfully applied to bond LM (Table 1; LM_8) to polymers containing aryl side groups by the Friedel-Crafts reaction. This method can be used for bonding LM to various phenyl-containing polymers (such as polymers and copolymers of styrene, α-methylstyrene, vinylbenzylic and 2-propenyl benzylic ethers) by varying the structure of the reagents of the type $A-CH_2X$ (A = anthryl, X = Cl, OH) and the reaction conditions (catalysts: $SnCl_4$ or $BF_3 \cdot OEt_2$; solvents: methylene chloride or o-dichlorobenzene; temperature: 25 to −78 °C)[104].

3.2.3 Combined Methods

The combined (two-stage) methods use the copolymerization of the main monomer containing the "labeling" amount of the comonomer (bearing a reactive group) with subsequent modification of the units of the main monomer or the comonomer. It is desirable to apply these methods when the single-stage synthesis of a polymer with LM is impossible for some reasons or very complicated or when it is necessary to introduce LM with different dynamic or optical parameters into the samples of one polymer.

For example, the single-stage preparation of the polymers and copolymers of styrene with LM of the 9-anthrylmethyloxycarbonyl structure (Table 1; LM$_1$) by copolymerization is accompanied by homolytic side reactions (Scheme 1). However, such labeled polymers can be obtained in two stages: by copolymerization of the main monomer with the "labeling" amount of acrylic, methacrylic or vinylbenzoic acids and subsequent anthrylmethylation of the carboxylic groups of the polymer using 9-anthryldiazomethane.

The treatment of a copolymer of methyl methacrylate with a "labeling" amount of methacrylic acid units by 9-anthryldiazomethane, 10-methyl-9-anthryldiazomethane or 10-phenyl-9-anthryldiazomethane serves as an example of the preparation of polymer (copolymer) samples of the same structure but with different LM by a combined method[105].

3.3 Methods for the Preparation of Polymers with Luminescent Markers of the Anthracene Structure in the Main Chain

The introduction of LM into the main chain of polymers is more difficult. For this purpose, a method based on free-radical copolymerization with the participation of 9-vinylanthracene (III) proceeding according to Scheme 2 has been developed.

The macromolecules formed in the copolymerization contain 9-alkylanthracene (IV) and 9-methylene-9,10-dihydroanthracene units (V). Under the usual conditions of the copolymerization of methyl methacrylate and III, the ratio of structures IV to V in the polymer varies from 1 : 15 to 1 : 50. If the copolymer is treated with trifluoracetic acid in a polar medium, units V undergo rearrangement, and LM units (Table 1; LM$_5$) of 9,10-dialkylanthracene structure (VI) appear in the main chain.

This method can be successfully used for polymers of the methacrylic series (the macroradicals of which do not react homolytically with the anthryl group). It can also be applied to introduce LM into the chains of such polymers as polystyrene. However, in this case the copolymerization can be complicated by homolytic reactions between the growing macroradicals and the anthracene groups (see Scheme 1). Presumably, it is possible to avoid side reactions by low-temperature cationic polymerization[77].

Another method of synthesis of polystyrene containing LM in the main chain is based on the reaction of "living" ends of the macromolecules obtained in the anionic polymerization of styrene with a bifunctional reagent, namely 9,10-dibromomethylanthracene[106].

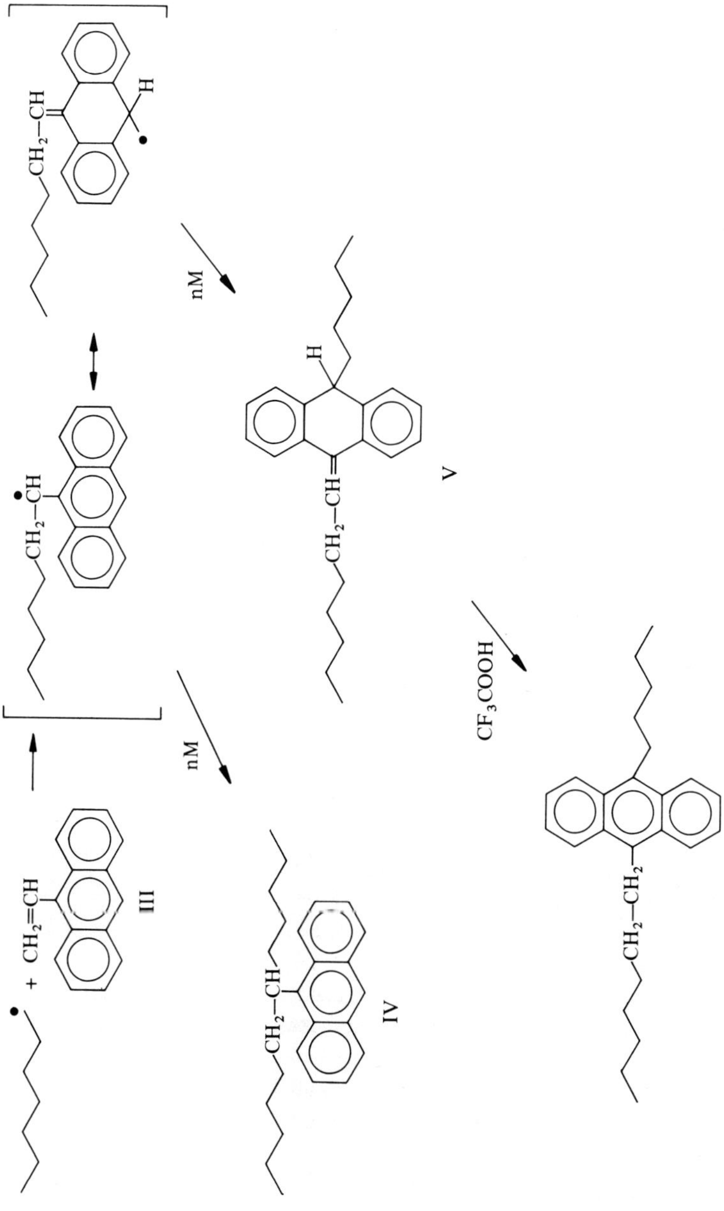

Scheme 2

3.4 Methods for the Preparation of Branched and Cross-Linked Polymers with Luminescent Markers

The investigation of cross-linked and branched polymers by the *PL* method permits to obtain information on the dynamic characteristics of both cross-linked junctions and linear fragments of the polymer system. Depending on the aims of the investigations, LM should be located either in a cross-linked junction or in a linear fragment of the polymer.

Polymers with LM in crosslinked or branched junctions can be synthesized by two methods:
1. The copolymerization of the main monofunctional monomer with a bifunctional monomer containing a luminescent group, e.g. with a dimethacrylic ester of 9,10-bis-(hydroxymethyl)anthracene[83].
2. The reaction of macromolecules with bifunctional anthracene-containing reagents, e.g. in the Friedel-Crafts reaction between the phenyl ring of polystyrene and 9,10-bis(chloromethyl)anthracene.

Cross-linked or branched polymers containing LM in the linear fragments of the system can be prepared by three methods:
1. Three-component copolymerization of the main monofunctional monomer and a monofunctional monomer bearing LM with a bifunctional monomer without a luminescent group,
2. cross-linking of linear macromolecules containing covalently bonded LM,
3. LM bound by the reaction between an anthracene-containing reagent and functional groups of finely dispersed swollen particles of a cross-linked polymer[48].

The methods of synthesis described in Chap. 3 were used to obtain labeled luminescent polymers investigated by the *PL* method.

4 Investigation of Polymer Solutions by Polarized Luminescence

4.1 Application of PL for Studying the Mobility of Various Parts of the Polymer Chain

The methods of the synthesis of labeled polymers with anthracene-containing LM (see Sect. 3) allow the preparation of polymers with LM of identical structure

Table 7. Relaxation times τ_w (ns) for conjugated macromolecules with the marker in the main chain (LM$_5$), in side groups (LM$_1$) or at chain ends (LM$_{10}$), 25 °C. Molecular weight of polymers M ~ 10^5

Polymer	τ_w (ns)		
	LM$_1$	LM$_5$	LM$_{10}$
PMMA in methyl acetate	3.9	8.3	3.2
PS in toluene	5.2	8.9	4.1
PMAA in water	77	120	–
PMAA in methanol	7.4	19	–

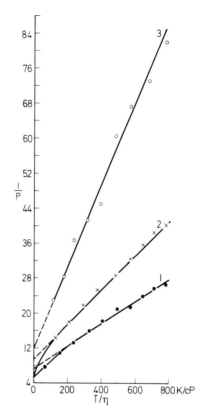

Fig. 2. Dependences of $\frac{1}{P}$ on $\frac{T}{\eta}$ at 25 °C for poly-
(methyl methacrylate) containing (1) LM_5 markers
in the main chain (2) LM_1 markers in side groups
and (3) LM_{10} at the chain end.

located in the main chain, in side groups or at the chain end. These experiments
permits not only to compare the mobility of the chain end and of the in-chain parts
(Table 7), but also to study the low-frequency relaxation processes in the side
chains[74]. Figure 2 shows the dependences $1/P = f(T/\eta)$ at 25 °C for PMMA in
methyl acetate with the LM of the anthracene structure in the main chain, LM_5, in
the side group, LM_1 and at chain end, LM_{10}. (For the structure of LM, see Table 1).

4.2 Influence of the Chemical Structure of the Polymer on the Intramolecular Mobility (IMM) of Macromolecules in Solution

The investigation of the IMM of polymers of various classes by PL [40, 48, 57, 100]
using anthracene-containing luminescent markers of similar structures, mainly
9-anthrylmethyloxycarbonyl groups, allows conclusions to be drawn on influence
of the substituents (their position, structure, and length) on the IMM of the polymer.

4.2.1 Polymers with a Heteroatom in the Main Chain

Polymers with a heteroatom ($-O-$) in the main chain exhibit the highest intramolecular mobility. This applies in the first place to polyoxyethylene ($\tau_{IMM} < 1$ ns).

Therefore, a precise determination of τ_{IMM} by the *PL* of anthracene-containing LM is impossible since the IMM of the polymer is higher than the mobility of the marker itself and, hence, the mobility of the latter becomes the predominat relaxation process.

Among polymers with a heteroatom in the main chain, polyamic acid is of interest being a representative of a wide class of new polymer materials with high thermal stability. The anthracene-containing markers could not be used for the investigation of the IMM of this polymer owing to a high conformational depolarization of luminescence[60]. A polyamic acid with luminescent units in the main chain has been synthesized[107]. The analysis of data on the IMM of polyamic acid in dimethylformamide obtained by the *PL* method reveals that the kinetic flexibility of its main chain is determined by the rotational mobility of the chain fragment located between the oxygen atoms. This mobility has been determined on the basis of theoretical considerations by Yu. E. Svetlov and Yu. Ya. Gotlib[107].

Hence, the theoretical calculations[108] and the conclusions on the free unhindered motion about the oxygen bridges in the main chain of polyamic acid have been confirmed.

4.2.2 Intramolecular Mobility of Vinyl Polymers

Among the vinyl polymers, polymethylene and poly(methyl acrylate) (PMA) exhibit the highest mobility ($\tau_{IMM} < 1$ ns). According to data obtained by other methods[109], τ_{IMM} of these polymers is $10^{-10} - 10^{-11}$ s. This means that the substituent $-CO-O-CH_3$ virtually does not hinder the rotation of the monomer unit in the main chain. The introduction of the methyl group in the α-position (PMMA) or the retention of the hydrogen atom at C_α but the replacement of $-CO-O-CH_3$ by a more bulky substituent, e.g. a phenyl ring as in polystyrene (PS) leads to a sharp increase in the intramolecular hindrance; $\tau_{w(IMM)}$ then becomes 6.0 ns (for PMMA in toluene) and 5.5 ns (for PS in toluene). This increase on passing from PMA to PS is due to the enhanced size of the side substituent and to its greater steric interactions with other groups of the main chain. The same phenomenon occurs on passing from PMA to PMMA. The introduction of the α-CH_3 group hinders the transfer motion of the monomer unit owing to strong steric interactions.

We will now consider the effect of the introduction at the β carbon atom of more bulky substituents than the hydrogen atom. To answer this question, polydimethoxyethylene (PDME) has been investigated[89]. The introduction of another even not very bulky substituent ($-OCH_3$) at the β carbon atom lowers the IMM of this polymer to the same value as those of PMMA and PS (Table 8). The same conclusion on the pronounced effect of the substituent at C_β can be drawn from the data obtained for poly(vinylbenzyl ether) (PVBE) and poly(propenylbenzyl ether) (PPBE) (Table 8). In contrast, the second substituent (such as the CH_3 group at the α-carbon) does not significantly affect the IMM of the polymer if the first substituent causes high steric hindrance to the transfer motion of the monomer unit about the $-C-C-$ bond of the main chain as in PS. In Table 8 are compliled data on PS and poly(α-methylstyrene), P(α-CH_3S), in toluene (τ_w = 5.1 and 8.6 ns, respectively).

Table 8. Comparison of $\tau_w(1)$ values for polymers with α and β substituents and of $\tau_w(2)$ for polymers of the corresponding structure containing hydrogen atoms at the α and β carbon atoms at 25 °C

Polymer	Solvent	$\tau_w(1)$ (ns)	$\tau_w(2)$ (ns)
PMMA	Methyl acetate	3.9	
PMA	Methyl acetate		<1
PMAA	Methanol-water (60:40)	6.5	
PAA	Methanol-water (60:40)		3.2
P(α-CH$_3$S)	Toluene	8.6	
PS	Toluene		5.1
PPBE[a]	Ethyl acetate	12.4	
PVBE[b]	Ethyl acetate		7.6
PDME	Methyl acetate	4.3	
see PMA	Methyl acetate		<1

[a]   PPBE polypropenylbenzyl
[b]   PVBE poly(vinylbenzyl ether)

4.2.3 Length of Side Chains and IMM of the Polymer

Let us consider the effect of the length of side substituents on the IMM of the polymer. The necessary data have been obtained[86, 87, 100] by using the *PL* method and anthrylacyloxymethane LM for poly(alkyl methacrylates) of the following structure:

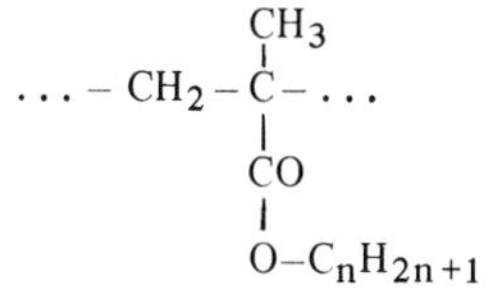

$$
\begin{array}{c}
\qquad \text{CH}_3 \\
\qquad | \\
\ldots - \text{CH}_2 - \text{C} - \ldots \\
\qquad | \\
\qquad \text{CO} \\
\qquad | \\
\qquad \text{O} - \text{C}_n\text{H}_{2n+1}
\end{array}
$$

for n = 1, 4, 6, 10, 16 and 22 and for cholesterol, cetyl and methyl esters of poly-(n-methacryloyl-ω-amino carboxylic acids) of the structure

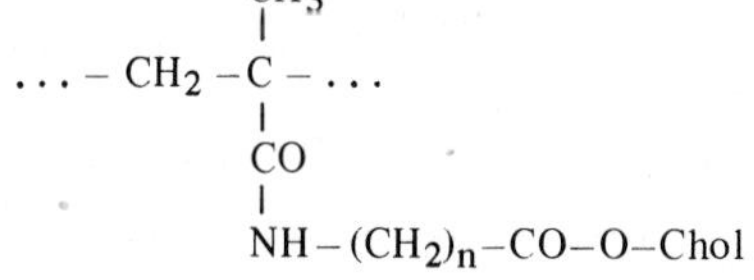

$$
\begin{array}{c}
\qquad \text{CH}_3 \\
\qquad | \\
\ldots - \text{CH}_2 - \text{C} - \ldots \\
\qquad | \\
\qquad \text{CO} \\
\qquad | \\
\qquad \text{NH} - (\text{CH}_2)_n - \text{CO} - \text{O} - \text{Chol}
\end{array}
$$

PChMAA-6, PChMAA-11, PCMAA-11 and PMMAA-11 (Table 9). When the polymer contains an α-methyl group, the increase in the length of the side substituent up to 22 carbon atoms has little influence on the IMM of the polymer in polar solvents for which the interactions between the side chains are weaker than in other solvents.

Table 9. Relaxation times τ_w^{red} (ns) for comb-like polymers in polar solvents at 25 °C (η_{red} 0.38 cP) PMA poly(alkyl methacrylates), n is the number of C-atoms in the alkyl chain

Polymer / Solvent	τ_w^{red} (ns) Chloroform	τ_w^{red} (ns) Dichlormethane
PMA-1	2.5	2.2
PMA-4	4.1	3.5
PMA-6	5.3	6.3
PMA-10	6.5	6.1
PMA-16	8.8	8.5
PMA-22	10.5	9.8
PMMAA-11	12	15
PCMAA-11	15	14
PChMAA-6	22	27
PChMAA-11	22	32

4.2.4 Intramolecular Mobility of Cellulose Derivatives

The *PL* method has been used to investigate the IMM of the macromolecules of carboxymethylcellulose (CMC)[48]. Anthrylacyloxymethane luminescent groups were used as LM. It was found that the presence of the cellobiose units in the main chain or the formation of intramolecular hydrogen bonds leads to a marked decrease in IMM. The value of τ_w for CMC in water is 57 ns. As Tables 11, 12 show, these high values of τ_w are characteristic of the IMM of macromolecules with the elements of the internal structure such as poly(methacrylic acid)(PMAA) in water. For non-ionized PMAA in water, τ_w is ca. 77 ns for LM in the side chain (9-anthrylmethyl-oxycarbonyl groups).

4.2.5 Influence of the Stereoregularity of the Polymer Structure on its Intramolecular Mobility

Comprehensive experimental material has been obtained on dielectric relaxation in solutions of polymers with various stereoregularities (e.g. PMMA)[110].

Higher sensitivity of the IMM of the polymer to the structure and contents of stereoregular sequences was detected by the *PL* method for a polymer bearing methoxy groups at each carbon atom of the main chain, namely poly-(1,2-dimethoxy-ethylene)[89]. The polymerization conditions were varied in such a manner as to obtain PDME samples with different stereoregularities[89] characterized by the contents of three meso-triads "mm" of the structure

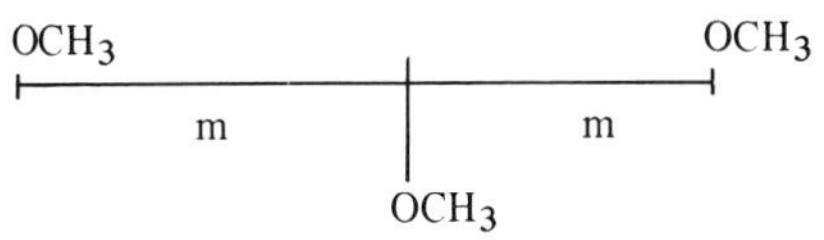

OCH₃ ——————|—————— OCH₃

        m   m

              OCH₃

The presence of these triads in the polymer and their contents (the fraction of R) were determined by NMR spectra[89]. The dependence of τ_w relaxation times determined by *PL* on R (fraction of triads) for PDME with anthracene-containing LM is very pronounced: τ_w = 10,3 ns for R = 74% and 4,5 ns for R = 86% (methyl acetate, 25 °C). This may be due to the presence of substitutions at each carbon atom in the chain.

4.2.6 Intramolecular Mobility of Random Copolymers

The *PL* method has been used to investigate the influence of the composition of random copolymers on the IMM of the polymer chain[48, 98, 99, 111]. Methyl acrylate and methyl methacrylate (the chain mobilities of the corresponding polymers differ by two orders of magnitude), styrene and α-methylstyrene (the relaxation parameters of the corresponding polymers are very similar) and, finally, styrene and methyl methacrylate were chosen as components of the random copolymers. Anthracene-containing LM were used for all these copolymers.

Data on the copolymers of styrene and α-methylstyrene and those of methyl acrylate (MA) and methyl methacrylate (MMA) are listed in Fig. 3. The extreme character of the dependence of IMM on copolymers of styrene and α-methylstyrene is quite understandable, if we recollect that the lower IMM of polystyrene chains than that of PMA may be associated not only with the large size of the phenyl ring but also with the specific interactions of the planes of neighboring phenyl rings[112]. The introduction of α-methylstyrene units prevents the mutual dispositions of the interacting phenyl rings and increases the IMM of the copolymer, if the content of α-methylstyrene units is low. At higher content of these units, the IMM of the copolymer approaches that of poly α-methylstyrene.

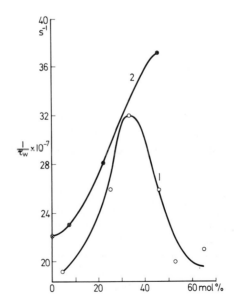

Fig. 3. Intramolecular mobility of copolymers of various compositions.
1 Styrene and α-methylstyrene copolymer in toluene, mole-% of α-methylstyrene;
2 Copolymer of methyl acrylate and methyl methacrylate in methyl acetate, mole-% of methyl acrylate; 25 °C

4.3 Influence of Intramolecular Hydrogen Bonds on the IMM of the Polymer

The IMM of the polymer decreases greatly, when intramolecular hydrogen bonds are formed in macromolecules. This effect becomes apparent, for example, if the data on PMMA and PMA and the corresponding acids are compared. The data on organic solvents should be compared (Table 10). In water the IMM of polyacids decreases markedly not only because intramolecular hydrogen bonds are formed but also for other reasons (formation of intramolecular structures in PMAA and hydrophobic interactions of non-polar groups). For aqueous solutions of PAA, the action of intramolecular hydrogen bonds on IMM can be seen, if the values of τ_{IMM} for non-ionized ($\alpha = 0$) and ionized PAA are compared (Table 10).

Table 10. Values of τ_w^{red} for acrylic esters and corresponding polyacids; $\eta_{red} = 0.38$ cP, 25 °C

Polymer	Solvent	τ_w^{red} (ns)
PMA	Methyl acetate	<1
PAA	Methanol	2.3
PMMA	Methyl acetate	3.9
PMAA	Methanol-water (60:40)	6.5
PAA, $\alpha = 0^a$	water, 0.02 M HCl	10
PAA, $\alpha = 0.1$	Water, NaOH	4.9

a α is the degree of ionization of carboxyl groups in the polyacid

4.4 Influence of Hydrophobic Interactions of Non-Polar Groups of Macromolecules in Water on the IMM of the Polymer

Data on some polymers containing non-polar groups in water and organic solvents or on their aqueous solution at various temperatures are reported as examples (Table 11). It should be borne in mind that, when aqueous solutions are heated to 60 °C, the hydrophobic interactions of non-polar groups (compounds) increase[113].

4.5 Intramolecular Mobility and Formation of Intramacromolecular Structures

Experiments reveal that the formation of the elements of various internal structures in macromolecules changes the IMM of the latter[114-117]. The appearance of local regions with internal structures in the molecules of PMAA[114,115], the coil-helix transition in synthetic polypeptides[101,116] and, finally, the formation of a globular structure in poly(1,2-dimethoxyethylene) molecules[117] lead to a several fold increase in the relaxation times characterizing IMM an, when the globular structure appears, the relaxation times increase by an order of magnitude (Table 11).

Table 11. Values of τ_w^{red} for solutions of polymers with nonpolar groups. η_{red} = 0.89 cP

Polymer	Solvent	τ_w^{red} (ns)
PAmA[a]	Dimethylformamide (DMF)	2.6
	Water (10% DMF)	130
PMAA	Water, 25 °C	77
	Water, 60 °C	120
PDME	Water, 25 °C	8.3
	Water, 63 °C	98

[a] PAmA – polyamic acid

4.5.1 Changes in the Internal Structure (of Local Regions with Internal Structure) in Macromolecules of Poly(methacrylic Acid) (PMAA) and IMM of the Polymer

The data in Fig. 4 show that the disappearance of the internal structure of PMAA during ionization of the carboxy groups of the polyacid[118, 119] is revealed not only by the cooperative increase in the intrinsic viscosity [η] of aqueous solutions of PMMA, but also by a considerable increase in its IMM that occurs cooperatively at the same degrees of ionization α at which the size of the macromolecular coil increases[114]. The internal structure of PMAA in water is stabilized by intramolecular hydrogen bonds and by the interaction of non-polar methyl groups. The introduction of additives weakening hydrophobic interactions breaks the internal structure of PMMA[118] and increases its IMM[115]. The greater the amount of non-polar groups in the molecule of the alcohol additive (e.g. methanol, ethanol, propanol) the greater this increase (Fig. 5). A comparison of the changes occurring in IMM of PMAA and poly(acrylic acid) (PAA) molecules (Fig. 4)[55, 114] during ionization of carboxy groups also shows that the contribution of the interactions of non-polar methyl groups to the change in the IMM of PMAA is very great. PMAA exhibits a greater

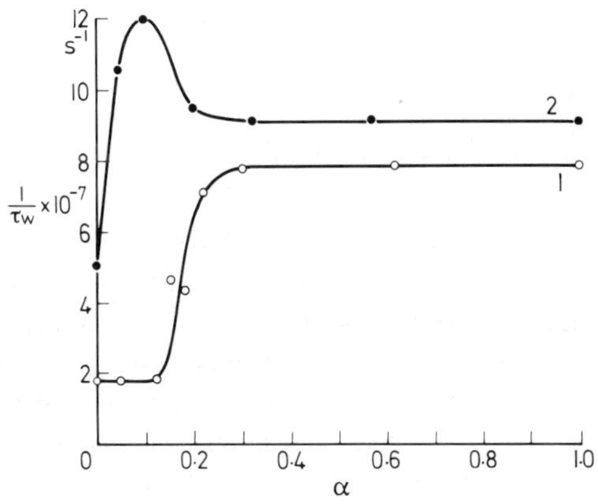

Fig. 4. Dependence of the IMM coefficient $\theta = \dfrac{1}{\tau_w}$ for poly(methacrylic acid) (1) and poly(acrylic acid) (2) acids on the degree of ionization of carboxy groups, α, at 25 °C. LM$_7$ marker

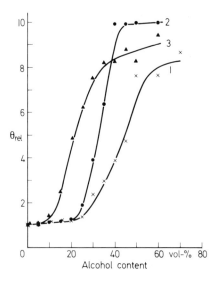

Fig. 5. Relative change in the IMM coefficient
$$\theta_{rel} = \frac{\theta\,(\%\ alcohol)}{\theta\,(water)}$$ for aqueous alcohol solutions of poly(methacrylic acid) (LM$_7$ marker) at 25 °C. 1 water-methanol, 2 water-ethanol, 3 water-propanol. (All the values of θ are reduced to the same viscosity of the solvent, $\eta_{red} = \eta_{water} = 0.89$ cP)

intramolecular hindrance than PAA (Tables 10, 11) and $1/\tau_w$ changes very little (Fig. 5) at low degrees of ionization of the carboxy groups or at low contents of the breaking agent: alcohol or dioxan. The presence of this region of constant $1/\tau_w$ in the dependence of IMM on the fraction of the agent-breaking intrachain contacts and the subsequent sharp (fourfold) change show that local parts with an internal structure are formed in non-ionized PMAA. These data are in good agreement with the results obtained by use of other methods (potentiometric titration, viscometry, etc.)[118].

4.5.2 Helix-Coil Transition in Synthetic Polypeptides and their IMM

The changes in IMM during the conformational helix-coil transformation in synthetic polypeptides have also been studied by PL [101]. The polymer used was poly(glutamic acid) (PGA) with anthracene-containing luminescent markers in the side chains (one marker per 1000 monomer units). The structure and the position of the marker are:

$$-CH_2-CH_2-\overset{\overset{\textstyle O}{\|}}{C}-O-CH_2-$$

The plots of $1/P$ vs. (T/η) at 25 °C for aqueous solutions of PGA at various pH in the range of the helix-coil transition are shown in Fig. 6. The lifetime of luminescence τ_f was measured with a phase fluorometer. The values of the parameter of high-frequency twisting vibrations of the marker $1/P_0'$ are listed in Table 4 and the dependence of the IMM of PGA on pH is shown by the plot of τ_w vs. pH (Fig. 7, curve 1).

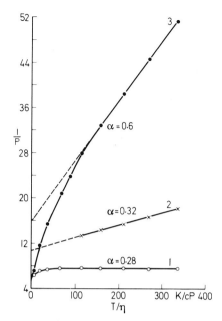

Fig. 6. Dependence of $1/P$ on T/η at 25 °C of poly(glutamic acid) (PGA) in water-sucrose solutions at various degrees of ionization α of carboxy groups of PGA; LM_3 marker 1 $\alpha = 0.28$; 2 $\alpha = 0.32$; 3 $\alpha = 0.6$

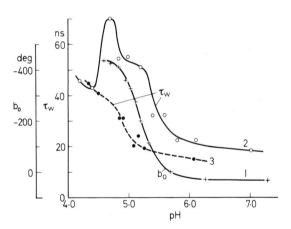

Fig. 7. Plot of τ_w vs. pH for aqueous solutions of poly(glutamic acid) (PGA) and its copolymers with leucine 3 PGA (96% Glu − 4% Leu), 2 (83% Glu − 17% Leu); (1) b_0 for PGA; 25 °C. The LM_3 marker is an anthrylacyloxymethane group at the end of the side chain of the Glu unit (1 LM per 1000 monomer units)

The change in the optical activity of PGA, the dependence of b_0 vs. pH (Fig. 7, curve 4) is an indication of the helix-coil transition. These data show that the breakdown of the α-helix in PGA is accompanied by a threefold increase in the IMM of the chain occurring cooperatively at the same pH at which b_0 (the fraction of α-helical sequences of PGA) changes.

There is another interesting consequence of the use of *PL* for the study of PGA with luminescent markers introduced into the carboxy groups of the polypeptide. High-frequency twisting vibrations of the marker located in these groups are very sensitive to changes in the interaction of the side chains of PGA taking place when the pH of the solution is varied (Table 4). These changes may result from the formation of hydrogen bounds between the carboxy groups in the side chains of PGA and their breaking during ionization. This interpretation is supported by the data obtained from PGA containing 9-antryl groups located only at the ends of the PGA chains. The structure and location of the luminescent group are as follows:

In this case the parameter $1/P_0' = 13$ and remains unchanged for PGA solutions at all pH values $(7.0 - 4.2)$[116].

A comparison of τ_w (pH) and τ_e (pH) in relative units (τ_e is the time describing the mobility of the chain end) suggests that during structure formation not only the mobility of the in-chain units of the polypeptide main chain changes but also the mobility of its end.

4.5.3 Coil-Globule Type Transition and IMM of Polymer

The high sensitivity of IMM to the processes of the formation of intramacromolecular structures allows to use the *PL* method for studying the IMM of the polymer under the conditions of its precipitation, i.e. at a high increase in polymer-polymer interactions[48, 117]. In this case the macromolecule can be precipitated or can form interchain contacts with neighboring macromolecules. To study the processes of the formation of intramacromolecular structures when the polymer-polymer interactions drastically increase, it is necessary to decrease the probability of the formation of interchain contacts. For this purpose, first, very dilute solutions with a polymer content as low as 0.001% should be studied and, second, the preferential formation of intrachain rather than interchain contacts should be ensured. This is possible for macromolecules the chemical structure of which does not prevent the formation of contacts existing for extended periods of time. The macromolecular contacts due to hydrophobic interactions of non-polar methyl groups in water are of long duration. Thus, by using the *PL* method it is possible to observe the intramacromolecular coil-globule transition and to study its dependences on structural parameters of the polymer[117]. For these studies the water-soluble poly(1,2-dimethoxyethylene) (PDME)[89], $-[CH(OCH_3)CH(OCH_3)]-$ was selected, i.e. the methyl group, required for prolonged intrachain contacts in water, is bonded to each carbon atom of the polymer chain. Moreover, the methyl group is bonded to the main chain through the ether oxygen, i.e. it is very mobile and thus all conditions exist for the different arrangements of chain segments bearing methyl groups in the formation of intrachain contacts. To make these contacts and hydrophobic interactions stronger[113], the PDME solution was heated from 25 to 60 °C. The concentration of PDME in water was 0.003%. Under these conditions, it is possible to exclude the formation of interchain polymer-polymer contacts[117]. This was checked by special experiments and confirmed by the absence of the concentration dependence of τ_w at high temperatures, when the concentration was changed by an order of magnitude (from 0.003 to 0.03%). The change in τ_w (Fig. 8) reflects the intramacromolecular conformational

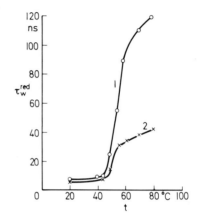

Fig. 8. Dependence of τ_w^{red} on temperature for poly-(dimethoxyethylene) (PDME) in water. Molecular weight of the polymer M: 1 80000, 2 25000, $\eta_{red} = 0.89$ cP

transition which makes the PDME coil more compact and reduces its size to that of a protein globule. The dependence of this transition on the molecular weight of the polymer (Fig. 8) also supports the conclusion that a transition of the coil-globule type occurs in PDME macromolecules upon heating in water. The *PL* method can also be used to reveal the increasing compactness of the PDME coil at such high dilution. As the size of the coil decreases, the value of τ_{wh} (time characterizing the rotational mobility of the coil as a whole) also decreases. The value of τ_{wh} becomes comparable with the lifetime of luminescence of the anthryl group in water, $\tau_f = 9$ ns. Hence, the new relaxation process is revealed in the parameters of *PL*, and the time of this process can be evaluated by use of Eq. (4.5.1)

$$1/\tau_w = 1/\tau_{wh} + 1/\tau_{IMM} \tag{4.5.1}$$

from the values of τ_w at the maximum of the temperature dependence $\tau_w(t)$. The maximum (inflection point) is related to the development of two opposite relaxation processes and is expressed by Eq. (4.5.2)

$$(1/\tau_{wh})^* = (1/\tau_{IMM})^* \tag{4.5.2}$$

valid for t^*, the temperature at which the dependence $\tau(t)$ attains the plateau region. Eq. (4.5.3)

$$\tau_{wh} = 1, 2 [\eta]\eta \, M/RT \tag{4.5.3}$$

is used to evaluate $[\eta]^*$ at 63 °C corresponding to a sharp increase in intramolecular hindrance and intrachain contacts. A comparison of the value of $[\eta]^*$ and that of $[\eta]$, determined by viscometry of the PDME solution at 25 °C, shows that the size of the PDME coil considerably decreases to that of a protein globule, $[\eta] = 0.03$ dl/g (Table 12).

Table 12. Relaxation times for PDME in water at 25 °C ($\tau_w^{25°}$) and at temperatures t* at the inflection point of the plot τ_w vs. t. Values of $[\eta]$ for solutions of this polymer were measured at 25 °C and calculated from the values of τ_w^* using Eq. (4.5.1). Values of relaxation times are reduced to the values of viscosity $\eta_{red} = 0.89$ cP. PDME I(II) and III are samples with different stereoregularity

	PDME I	PDME II	PDME III
$\tau_w^{25°}$ (ns)	8.3	7.5	18
τ_w^{*red} (ns)	98	30	41
$[\eta]^{25°}\left(\dfrac{dl}{g}\right)$	0.62	0.25	0.23
$[\eta]^*\left(\dfrac{dl}{g}\right)$	0.03	0.03	0.05
M	80000	25000	18000

The coil-globule transition can also be observed by *PL* (from changes in the IMM of the polymer), when polymers are precipitated in organic solvents. These data were obtained for PMMA in toluene precipitated with octane or other non-solvents. Figure 9 shows the dependence τ_w on the concentration of octane used as the precipitant for PMMA fractions of various molecular weights. Naturally, for PMMA in

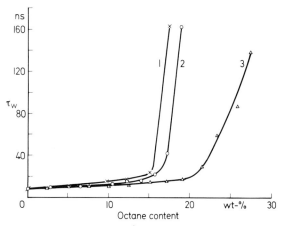

Fig. 9. Relaxation times τ_w^{red} for solutions of poly(methyl methacrylate) (PMM) (LM_5 marker) of various molecular weights in toluene-octane mixtures as a function of the octane content at 25 °C, $\eta_{red} = 0.38$ cP. M: 1 6.6×10^5, 2 3×10^5, 3 6×10^4; the polymer concentration in solution is 0.003%

organic solvents it is impossible to observe the increase in the coil compactness on the monomolecular level. The reasons for this were discussed at the beginning of this section.

High sensitivity of the IMM of the polymer to the processes of the formation of intramacromolecular structure and the possibility of using the *PL* method in the investigation of highly dilute solutions of polymers has allowed to utilize it for the solution of topical problems of the physics of synthetic macromolecules modelling protein globules[101] or molecules bearing mesogenic groups and forming polymer systems with liquid-crystalline properties[87, 100].

4.5.4 Coil-Compact (Globular) Transition in Random Copolymers of Polypeptides and their IMM

Synthetic copolymers of glutamic acid and leucine (Glu-Leu) of various compositions have been used to solve the problem of whether the variation in the composition of a two-component random copolymer consisting of peptide units can give rise to the formation of a compact globule modelling a protein globule[101]. Although this problem was solved. 4% of leucine units in the Glu-Leu copolymer (96 : 4) were found to be insufficient for changing the conformational properties of the copolymer as compared to a polypeptide chain consisting only of Glu units. However, 17% of Leu in the copolymer with 83% Glu lead to the formation of a compact globular structure in aqueous solutions of non-ionized macromolecules of the copolymer as shown by the considerable increase in intramolecular hindrance and in the compactness (down to the size of a protein globule) of the coil (Fig. 7). This increase in the compactness of macromolecules and in their rotational mobility is indicated by a distinct maximum of the plot of τ_w vs. pH (Fig. 7, curve 3).

4.5.5 Formation of Mesophase Nuclei in Polymers with Mesogenic Groups and IMM of the Polymers

The problem of the role played by the mesogenic groups in the changes in intra- and intermolecular contacts, in other words in the formation of supermolecular structures in solutions, is of great interest because its elucidation is closely related to the possibility of changing liquid-crystalline properties of the polymer system. The *PL* method has been used to investigate the relaxation properties of polymers with mesogenic groups in side chains in solution in order to establish the conditions of the formation of mesophase nuclei on a molecular level[87, 100]. Dilute (0.02%) solutions of cholesterol-containing polymers and copolymers (with butyl methacrylate) of various compositions were investigated: cholesterol esters of poly(N-methacryloyl-ω-aminocarboxylic acids) of the following structure:

$$H_3C-\underset{\underset{CH_2}{|}}{\overset{|}{C}}-CONH-(CH_2)_n-\overset{\overset{O}{\|}}{C}-O-Chol$$

where n = 6 and 11, PChMAA-6 and PChMAA-11, with anthrylacyloxymethyl luminescent groups (the marker is LM_1, one LM per 300 monomer units). It was found that
internal structures I and II are formed successively in PChMAA-11 at 25 °C on passing from a polar solvent, chloroform, to non-polar solvents like toluene (structure I)
and heptane (structure II). Structure I, including only the main chain of the polymer,
is formed in toluene. Structure II, including both the main chain and the side chains
of the comb-like polymer, is generated in heptane. Under the conditions of structure I,
the mobility of the main chain of PChMAA-11 greatly decreases: τ_w^{red} = 22 ns in $CHCl_3$
and τ_w^{red} = 100 ns in toluene (η_{red} = 0.38 cP) and the side chains (sc.) remain mobile
($\tau_{s.c.}^{red}$ is 0.6 ns in $ChCl_3$ and 3.3 ns in toluene, marker at the end of the side chain).
In structure II in heptane both the main chain and the side chains are hindered: for
PChMAA-11 $\tau_{s.c.}^{red}$ is greater than 30 ns. The nuclei of the mesophase appear under
conditions of structure II, when the mobility of these mesogenic groups at the ends
of hindered side chains decreases. Their formation changes the optical activity of the
solution of cholesterol-containing polymers (Fig. 10) and greatly decreases the IMM
of the main chain of PChMAA-11: τ_w^{red} increases up to 490 ns[87, 100]. If the temperature of the PChMAA-11 solution in heptane is below 30 °C, the interactions of
the nuclei of the mesophase lead not only to a cooperative increase in the intramolecular hindrance (Fig. 10) but also to a compactness of the coil which is related
to a decrease in τ_w, when the PChMAA-11 solution in heptane is cooled below 20 °C.
The nuclei of the mesophase are not observed in copolymers of PChMAA-11 with
butyl methacrylate in heptane, when the content of mesogenic groups in the copolymer is lower than 50 mol%[87, 100]. The IMM of the copolymer remains unchanged in
the temperature range from 70 to 10 °C. Hence, the probability of the formation of

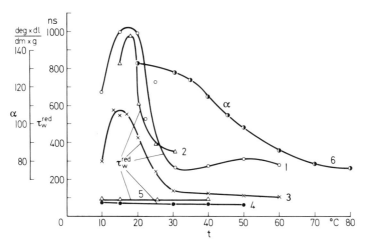

Fig. 10. Relaxation times τ_w characterizing the IMM of PChMAA-11 (1), PCMAA-11 (5),
copolymers P (ChMAA-11-MA-4) containing 10 (2), 25 (3) and 60 (4) mol% of MA-4-butyl
methacrylate units in aliphatic hyrocarbons as a function of temperature. The LM_1 marker is
the 9-anthryl methyloxycarbonyl group (one group per 300–600 monomer units); [α] is the
specific optical rotation for (1) in heptane. Values of relaxation times at various temperatures
are reduced to one solvent viscosity 0.38 cP

the mesophase nuclei depends not only on the temperature but also on the density of packing of mesogenic groups. These data imply that the mobility of these groups is the determining factor in the formation of mesophase nuclei.

4.6 Investigation of Block Copolymers (IMM of Polymer Blocks) by Polarized Luminescence

Block copolymers are an example of multicomponent polymer systems for which the problem of separate investigation of each component (block) of the copolymer is of particular interest. The *PL* method permits a successive separation of each block of the copolymer for the investigation of the IMM of the latter using luminescent markers[120]. The *PL* method was applied to the study of the three-block copolymers of the ABA and BAB types where A and B are PMMA and polystyrene(PS)blocks. The properties of the three-block copolymer PMAA (poly(methacrylic acid)-PS-PMAA with a hydrophilic PMAA block (LM$_1$ marker) are also of interest. In a three-block copolymer PMAA-PS-PMAA in water, the local sequences with internal structure in the PMAA are partially destroyed as compared to the PMAA homopolymer. The IMM of the PMAA blocks in water is higher (τ_w = 48 ns) than that of the PMAA homopolymer (τ_w = 77 ns, 25 °C). This increase may be due to the interblock hydrophobic interactions of non-polar groups in the PS and PMAA block of the three-block copolymer in water. The use of *PL* for the investigation of three-block copolymers in selective solvents allows to determine the conditions of the formation of various intramolecular structures — globular, ring, or dumbell structures — and to characterize the dependence of their formation on the chemical structure of block copolymers and the block sequence. A globular structure is formed in the precipitated block, if the solvent is thermodynamically suitable for the carrier-block irrespective of the position of the precipitated block. The three-block copolymer acquires a dumbell conformation, if the side blocks are precipitated and the solvent is thermodynamically suitable for the central block. The ring conformation appears in a three-block copolymer, if the side blocks are precipitated under Θ conditions for the carrier block of under the conditions prevailing at low (4 x 10^4) molecular weight of the carrier block. The data in Table 13 show the changes in the IMM of all blocks of the three-block polymer.

4.7 Formation of Intramolecular Structures in Polyelectrolytes in the Interaction with Surfactants in Water

An intramolecular structure is also formed in polymers when these interact with surfactants (S)[121] (Table 14). These changes in polymers by the action of S may be utilized for studies on intermolecular interactions in the polymer-S system. Only the polymer with bound S ions exhibits a lower IMM than does the original polymer. On the other hand, only those surfactants, that change the IMM of the polymer, come into prolonged ($>10^{-8}$ s) contact with the polymer[121]. The *PL* method was applied to a comparative investigation of the IMM of a copolymer of vinylamine (10%) with

Table 13. Relaxation times characterizing the intramolecular mobility of the polymer chain, τ_w^{red}, for the homopolymer and the block of the corresponding chemical structure and molecular weight M in three-block copolymers (PMMA-PS-PMMA) and (PS-PMMA-PS) at 25 °C; η_{red} = 0.38 cP. a) PMMA*, b) PS*. The polymer concentration in solution is 0.005%; the component under investigation with a luminescent marker is denoted by (*)

Polymer	a) PMMA* τ_w^{red} (ns)									
	Octane in toluene (wt-%) (PMMA is precipitated)						Acetonitrile in methyl ethyl-ketone (PS is precipitated)			
	0	12	17	22	40	50	0	35	50	85
PMMA* M = 130000	4.3	6.2	10.5	51			3.1	–	4.5	6.2
PS-PMMA*-PS 30000-170000-30000	4.3	6.2	10.5	51	Precipitant	Precipitant	3.1	–	4.9	15
PMMA*-PS-PMMA* 130000-36000-130000	4.3	6.2	33	82			3.1	–	4.5	6.2
b) PS*										
PS* M = 57000	3.6	3.9	4.1	4.2	4.7	4.8	3.0	6.0	55	200
PMMA-PS*-PMMA 60000-42000-60000	3.6	4.6	4.8	5.9	18	125	3.0	6.0	30	–

Table 14. Relaxation times τ_w for the VP-VNHR copolymers in aqueous solutions of octyl sulfate and dodecyl sulfate (The values of $\beta = \dfrac{[AS]}{[BNHR]}$ correspond to maximum changes in the IMM of the copolymer by the effect of alkyl sulfate; n is the number of carbon atoms in the alkyl side chain R (VP-VNHR) (90–10%)

AS	n				
	0	4	9	12	16
	τ_w (ns)				
C_8	33	46	84	110	160
C_{12}	106	120	130	180	250

vinylpyrrolidone (90%): $VNH_2 - VP$ (10:90) in the presence of S: alkyl sulfates (AS) with the structure $C_m H_{2m+1} OSO_3 Na$ where m = 8, 10, 12, 14 or 16 in water at various ratios

$$\beta = \frac{[S]}{[\text{polymer}]} \quad \text{(in moles per moles of the monomer units)}$$

and the IMM of alkylated copolymers with the structure VNHR-VP where
$R = C_n H_{2n+1}$ for $n = 8$, 10, 12 and 16 (Table 14). Groups exhibiting the structure

NHSO$_2$—(naphthalene)—N(CH$_3$)$_2$

were used as luminescent markers (one group per 120 copolymer units).

The data in Table 14 show that the IMM of the (VNH$_2$-VP) and (VNHR-VP) copolymers decreases in the interactions of these copolymers with alkyl sulfates (AS). The longer the alkyl substituent in AS or in the VNHR-VP copolymer, the stronger this decrease.

The data compiled in Table 15 reveal that the intramolecular hindrance observed in water with these systems is due to hydrophobic interactions of alkyl substituents. In methanol, the increase in the length of the alkyl substituent changes the IMM of the polymer only in accordance with the steric interactions of its side chains. The interactions of alkyl substituents in water also result in an increase of the compactness of the VNHR-VP macromolecules when the substituent is very long and in a decrease of $[\eta]$ of the copolymer solution (Table 15).

When the surfactant content in aqueous solutions of the VNH$_2$VP copolymer is low ($\beta = 1$), the intramolecular hindrance reaches a maximum. At a further increase in the content of S (dodecyl sulfate), the formation of S associates in solution and the transition of S ions from the polymer into solution becomes the dominant pro-

Table 15. Dependence of relaxation times τ_w^{red} (IMM) of the copolymers VP-VNHR ($R = C_n H_{2n+1}$) in water, aqueous salt solutions and methanol and intrinsic viscosity $[\eta]$ of the copolymer solutions in 0.5 M KCl at 25 °C on the length of alkyl substituent R. The values of τ_w are given for the viscosity of water, $\eta_{red} = 0.89$ cP

solvent	n				
	0	4	8	12	16
	τ_w^{red} (ns)				
Water	20	20	22	67	126
0.25 M NaCl	20	–	24	85	147
Methanol	17	–	–	–	24
	$[\eta]$ (dl/g)				
0.5 M KCl	0.30	0.30	0.29	0.14	–

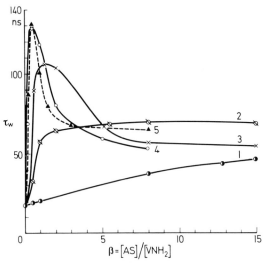

Fig. 11. Dependence of τ_w on the molar content of AS or $\beta = \dfrac{[AS]}{[VNH_2]}$ for the VP-VNH$_2$ copolymer (90%−10%) and AS with various length of alkyl chain $R' : R' = C_mH_{2m+1}$, m = 8 (1), 10 (2), 12 (3), 14 (4) and 15 (5). Polymer concentration is 0.5 mg/ml

cess because the interaction of the polymer with the added S ions and the formation of S associates. After S ions have passed into solution, the IMM of the polymer increases and relaxation times τ_w decrease (Fig. 11). The same phenomenon occurs if another polymer (II), a stronger competitor for the interaction with S, is introduced into the solution containing polymer I-S ($\beta = 1$)[122]. Hence, the *PL* method permits to study the competing interactions in polymer I-S-polymer II solutions. Naturally, this study requires the determination of the IMM of each copolymer in the presence of the other. This separation and investigation are possible by application of luminescent markers and the *PL* method.

4.8 Formation of Intramolecular Structures in Carboxy-Containing Polymers in Water by the Effect of Rare-Earth Ions

Ions of rare earth elements (REE) can form coordination bonds with the C=O groups of the macromolecules[123]. In polymers containing carboxy groups (PMAA, PAA, poly(glutamic acid)), the IMM of the polymer changes considerably due to the effect of the REE ions[124] (Table 16). This fact permits to study the efficiency of REE ions as a cross-linking agent of the polymer and the influence of solvent composition (water-methanol, water-dioxan) and ionization of carboxy groups on intermolecular interactions of these groups with REE ions (Table 16). The luminescent marker in carboxy-containing polymers is the 9-antrylmethyloxycarbonyl group, LM$_1$.

4.9 Investigation of Cross-Linked Polymer Systems by *PL*

Both polyelectrolyte networks[49, 51] and networks of other chemical structures are widely used owing to their peculiar properties (high effectiveness of bonding of

Table 16. Effect of $Tb^{3\pm}$ ions on τ_w^{red} of polyacids (PA) in various solvents (Values of τ_w are reduced to the viscosity of water, $\eta_{red} = 0.89$ cP; at $25°$)

Polymer	Solvent	τ_w^{red} (ns)	
		without REE	$\dfrac{[REE]}{[PA]} = 0.1$
PMAA	water, $\alpha = 0$[a]	77	320
	water + NaOH, $\alpha = 0.6$	13	34
	water-CH_3OH, 50:50	12	84
PAA	water, $\alpha = 0.05$	22	200
PGA	water, pH 6.8	15	105

[a] α is the degree of ionization

macromolecular ions, high strength of derived materials etc.). The relationship between the mobility of junctions and fragments of the network and the condition of its formation have been studied by the *PL* method. To study separately the mobility of these junctions and fragments, it is necessary to develop methods of covalent bonding of luminescent markers to linear fragments of the network and to use luminescent compounds as cross-linking agents. The *PL* method can be used for the determination of the local mobility of polymer networks with anthracene-containing luminescent markers if the polymer network is dispersed into particles of the 0.5 μm[49]. The *PL* was also used to study the local mobility of soluble polymer systems with covalent junctions, e.g. of PMMA of molecular weights of 5 x 10^4 or 5 x 10^5 with one junction per macromolecule. The mobility of the luminescent junction, formed by copolymerization of a bifunctional anthracene-containing monomer with methyl methylacrylate in solvents of various thermodynamic strength, was investigated. The data characterizing the mobility of the junction and that of the main chain are compared in Table 17. It was found that even the mobility of a single junction in a macromolecule is much lower than that of the inner part of the main

Table 17. Relaxation times τ_w^{red} for PMMA with covalent luminescent junctions in various solvents at $25\,°C$ ($\eta_{red} = 0.38$ cP); M = molecular weight of the linear fragment

Solvent fragment	τ_w^{red} (ns)			$[\eta] \left(\dfrac{dl}{g}\right)$ for linear PMMA (M = 28 x 10^4) in solution
	Linear PMMA	PMMA with covalent junctions		
	M = 28 x 10^4	M = 5 x 10^4	M = 50 x 10^4	
Benzyl acetate	–	9.1	12.3	0.89
Toluene	7.1	–	17.6	0.72
Methyl acetate	8.3	14.5	19.6	0.68
Butyl acetate	15.3	20.2	35.4	0.32

PMMA chain. This shows that the local density of the units near the junction increases. The influence of the increase in the local density of units on the mobility
of the junction becomes more pronounced as the thermodynamic strength of the
solvent for PMMA decreases and as the molecular weight of the polymer chain increases (Table 17).

Carboxylic cross-linked polyelectrolytes in water have been investigated by the
PL method[49]. 9-Anthrylmethyloxycarbonyl groups were also used as luminescent
markers in the amount of one group per 300 units of a cross-linked polyelectrolyte
based on methacrylic acid and ethylene dimethacrylamide (2.5 mol%). The problem
of the change in the mobility and conformational lability of cross-linked polyelectrolytes during ionization is of the greatest interest, since these properties determine
the performance of the polyelectrolyte network. The changes in the mobility of the
carboxylic polyelectrolyte network in water during ionization are compared with
the changes in IMM of linear PMAA in water during ionization (Fig. 12). The data
reveal a strong intramolecular hindrance in the cross-linked system as compared to
linear PMAA at all degrees of ionization and a strong (five fold) increase in IMM
during the ionization of carboxy groups. These changes occur cooperatively over a
narrow range of α — the degree of ionization of carboxy groups. In linear PMAA,
these variations are related to the existence of an inner structure in non-ionized
macromolecules[118] (see Sect. 4.5) and its disappearance during ionization[114, 118].
A similar character of changes in the IMM of cross-linked PMAA shows that, in linear
fragments of the PMAA polymer network at $\alpha = 0$, local sequences with an inner
structure also appear, but the cooperative increase in IMM determined by their de-

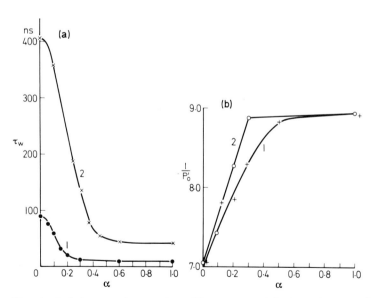

Fig. 12 a, b. Relaxation time τ_w (a) and PL parameter $1/P'_0$ (b) as a function of α for linear
soluble PMAA in water (1) and for cross-linked PMAA-C dispersed in water (2) at 25 °C. α is
the degree of ionization. The concentration of linear and cross-linked PMAA is 0.02%. The marker is an anthrylacyloxymethyl group (one group per 1000 monomer units)

struction occurs at higher α than in linear PMAA (Fig. 12a)[49]. It can also be seen that the dispersion of the polymer network in the solvent does not affect the *PL* parameters of the marker. Thus, the parameter $1/P_0'$ related to high-frequency torsional vibrations of the marker has the same value as in linear PMAA dissolved in water. Moreover, the coincidence in this parameter is retained when all carboxy groups are ionized (Fig. 12b). The *PL* method is also sensitive to the peculiarities of the structure of the soluble branched macromolecules obtained by polymerization in various media[125].

4.10 Use of the *PL* Method in the Study of Polymer-Polymer Complexes

Intermacromolecular hydrogen bonds, interactions of non-polar groups in water, and electrostatic interactions between macromolecules lead to the formation of polymer complexes consisting of cooperatively bonded continuous sequences of monomer units of complementary polymer chains or to the formation of poly-salts[126-133].

High sensitivity of the IMM of the polymer to changes in intramacromolecular interactions of various types (specific or Van der Waals interactions, etc.) permits to make use of the relaxation properties of the polymer for studying intermacromolecular interactions in polymer-polymer complexes (*PC*). A comparative investigation of the IMM of macromolecules constituting *PC* and single macromolecules of each of its components has been carried out for a number of *PC*[48, 131-133]. The *PL* method permits to study the IMM of single *PC* components, because of the possibility of using labeled or unlabeled macromolecules. The superscript[*] designates the labeled component of *PC*. Thus, in the (PMAA*-PEG) complex the PMAA macromolecules and in the (PMAA-PEG*) complex the poly(ethylene glycol) macromolecules are labeled. Anthracene-containing LM were also employed to study *PC*.

The effect of intermacromolecular interactions on the IMM of the polymer chains forming *PC* has been investigated in detail for PMAA-PEG complexes. Data

Table 18. Relaxation times τ_w for aqueous solutions of PMAA*-PEG (PDME) and PEG* (PDME*)-PMAA complexes for $\dfrac{[PEG(PDME)]}{[PMAA]} = 1$ (in moles of the monomer unit) and for uncomplexed PMAA and PEG(PDME); 0.1% PMAA; 25°C

System	τ_w(ns)	System	τ_w(ns)
PMAA	77	PEG	<1
(PMAA*-PEG) complex	290	(PEG*-PMAA) complex	350
(PMAA*-PDME) complex	330	(PDME*-PMAA) complex	300
		PDME	8.3

Table 19. Relaxation times τ_w for PAA and PMAA in polymer complexes with insulin formed at various pH and for uncomplexed PMAA and PAA molecules in an ammonium buffer; α is the degree of ionization; 0.1% PMAA, pH = 9.1, 25 °C

System	pH at which the complex was formed	τ_w(ns)
PMAA, $\alpha = 1$ M = 52000	–	13
PMAA-insulin complex	7.1	20
	8.2	83
	9.1	106
	10.6	28
PAA, $\alpha = 1$ M = 150000	–	11
PAA-insulin complex	9.1	20

on these systems have also been obtained by other methods[125–130]. Data in Table 18 reveal the changes occurring in the IMM of PMAA and PEG macromolecules owing to intermacromolecular interactions in the PMAA-PEG complex[48]. Similar changes in the IMM of polymers take place, when a *PC* of PMAA with poly(1,2-dimethoxy-ethylene)(PDME) is formed in water (Table 18)[48,131].

Data in Table 19 show the variations in the IMM of PMAA molecules (or poly-(acrylic acid), PAA), when these interact with insulin[132]. The polymer complex was separated from the unreacted molecules. The great decrease in intramolecular motions in PMAA macromolecules as compared to those of PAA in polymer complexes with insulin may be due to hydrophobic interactions between methyl groups of PMAA and non-polar groups of insulin in aqueous PMAA-insulin solutions. Kinetic characteristics of intermolecular interactions in the polymer-polymer complexes have also been studied by the *PL* method[133]. The dependence of kinetic parameters of intermolecular interactions on the structure of interacting chains, their length and the chemical nature of bonds in *PC* has also been investigated[133].

It is possible to determine the lifetime of intermolecular contacts in *PC* from the rate of equilibration in an exchange reaction of type I.

$$PC(P_1 P_2) + P_2^* \rightleftarrows PC(P_1 P_2^*) + P_2 \tag{I}$$

where the P_2^* macromolecules are identical with P_2 in their chemical structure but contains a luminescent marker. The *PL* method allows to investigate the kinetics of reaction (I) in solution in the following way: A certain amount of PMAA* is added to a mixture of unlabeled PMAA and PEG molecules forming *PC* in water. The change in the polarization of luminescence of the solution and the calibrating dependences obtained previously are utilized for the evaluation of the fractions of PEG molecules that passed from *PC* to PMAA* during different time intervals. In this way, the parameters characterizing the conversion of exchange reaction (I) were

Table 20. Dependence of the conversion of exchange reactions (κ parameter) on the chemical structure of interacting molecules and reaction time (in min) at pH 4.0 and 25 °C

Components of *PC*		κ		
matrix	oligomer	reaction time (min)		
		10	100	1000
PMAA $M = 2.5 \times 10^5$	PEG $M = 2 \times 10^4$	0.25	0.40	0.45
PAA $M = 2 \times 10^5$	PEG $M = 2 \times 10^4$	1.0	1.0	1.0
PMAA $M = 2.5 \times 10^5$	PVPD $M = 2 \times 10^4$	0.35	0.53	0.55
PAA $M = 2 \times 10^5$	PVPD $M = 2 \times 10^4$	0.40	0.65	0.85

determined for the following systems: PMAA-PEG, PAA-PEG, PMAA-polyvinyl-pyrolidone (PVPD) and PAA-PVPD (Table 20). The investigation of the kinetic lability of polymer-polymer complexes (*PC*) or the lifetime of intermolecular contacts stabilizing *PC* is a promising trend of the *PL* method.

5 Dynamic Theory of the Motions in Macromolecules and Polarized Luminescence

In the following sections, special attention will be paid to such aspects of the theory that are needed and used for a better understanding and prediction of the main features of the polarized luminescence of labeled macromolecules.

5.1 Macromolecules as a Dynamic Cooperative System

Long-chain macromolecules are dynamic systems consisting of many particles; hence, a broad spectrum of motions is possible on various spatial and time scales. The parameters of these motions exhibit various dependences on the temperature and viscosity of the solvent and various sensitivities to conformational changes.

First, rotatory and translational diffusion of the macromolecule as a whole occurs. Furthermore, a spectrum of intramolecular motions exists; they may be divided into relatively large-scale motions accompanied by a change in the size and shape of the whole macromolecule and small-scale local motions localized within the range of one or several units. When the dependence of local relaxation properties of polymer homologs on the molecular mass (which is observed sometimes) is considered, it should be borne in mind that some parameters of intramolecular organization, such as the average concentration of units in a coil or their local concentration in the vicinity of a given chain unit[47, 134, 135], can change with the molecular mass of the polymer. This feature is particularly reflected in the behavior of macromolecules either near the Flory Θ-point (in ideal solvents) or in poor solvents (precipitants).

The parameters of the rotatory and translational diffusion of the macromolecule as a whole and those of intramolecular large-scale motions depend on the size and shape of the macromolecule, on its hydrodynamic permeability and on the possible solvation of the solvent. (It is known that the latter factor should be taken into account in the study of compact protein globules in water.)

In contrast, the parameters of local motions are very sensitive to the local conformational microstructure of the polymer chain and to the interactions of units located at a large distance apart along the chain contour but close to each other in space (kinetic volume effects). The parameters of local motions also depend on the "external" viscosity of the solvent and "internal" viscosity of the polymer chain[136,137].

One of the major problems of the theory of intramolecular local mobility or kinetic flexibility lies in the establishment of the principal mechanism of mobility for the macromolecules of a given chemical structure and constitution and of local conformational microstructure. The question arises, which of the elementary acts leads to the ability of individual parts of the macromolecule for changing their shape during definite time intervals characteristic of a certain class of motions[46, 138−148].

Just as the equilibrium conformational properties of macromolecules, the theory of which has been developed in well-known classical works by Kuhn, Flory, Volkenstein and others[149−153], the kinetic properties of polymer chains can be determined by two main mechanisms of intramolecular mobility. First, it is the discrete rotational isomeric (rotameric) mechanism of mobility caused by the jump of small-chain segments (kinetic units) from certain energically stable allowed conformers into others[46, 140−145, 154−165]. Second it is the continuous mechanism of motion determined by the accumulation of vibrational displacements of units owing to the twisting or bending vibrations (and librations) or valence angles and angles of internal rotation[144,145, 166, 167].

Finally, a combination of the two limiting types of mechanisms of mobility is possible and even quite probable[144,145]. In this case, more complex types of cooperative motions occur along the most energetically advantageous paths in the multidimensional conformational space[168, 169].

In theory, each mechanism of mobility corresponds to the proper simplified "solvable" dynamic models of the polymer chain. They describe continuous viscoelastic[39, 93, 107, 170, 171] or discrete rotameric mechanisms of internal mobility[144, 156, 164, 165, 171−175]. Dynamic models, that can be mathematically treated, are often more simplified than the hypothetical models used in the theory of equilibrium properties.

For this reason, computer simulation methods (Monte Carlo and molecular and Brownian dynamic methods) have been developed not only for solving the problems of polymer statistics, but also for the investigation of the dynamic properties and the intramolecular mobility of polymers.

5.2 Dynamic Polymer Effects in Polarized Luminescence

The rotation of the oscillator in a luminescent marker inserted into the polymer chain is always accompanied by a complex translational and rotational motion of chain segments adjoining it. The main task of the theory is to take into account the polymer effects proper in PL. This problem has been considered by several authors including the authors of this review.

The theory describing the effect of various factors of short-range interactions along the chain on local relaxation properties is more developed and may help to explain the experimental data.

Experimental investigation of the PL of labeled macromolecules in solvents of various thermodynamic strengths (good or Θ-solvents) reveal that IMM strongly depends on the extent of swelling of macromolecules[40, 48, 56, 111]. These facts and similar results obtained in solutions of higher concentration by the dielectric relaxation method[59, 176, 177] and NMR[178, 179] suggest that the volume contraction of the coil and intermolecular contacts and collisions between units located at a certain distance from each other affect the local forms of mobility. The theory taking into account the effect of the amount and strength of intermolecular contacts on relaxation properties is elaborated at present[173, 180–184].

The dependence of $1/P$ (or Y) on T/η at low values of T/η observed experimentally[40, 48, 57] permits the investigation of high-frequency twisting or bending motions for which the characteristic times (periods and decay times) are small as compared to τ_f. To evaluate the amplitudes of these motions, a theory of high-frequency motions compatible with frozen chain conformation (when the times of segmental motion are still long as compared to τ_f) was developed[58] by Anufrieva, Gotlib et al.

The majority of existing theories can neither determine precisely the size and the microstructure of kinetic units in a polymer chain of a given chemical structure nor rigorously predict the mechanisms and the kinetics of conformational transitions. In these theories, the properties of kinetic units are postulated and the aim of the theory is to study the effects resulting from the linking of these units into the chain.

Polymer chains exhibit a considerable anisotropy of relaxation properties. As Stockmayer has pointed out[159] and other authors have investigated in detail[46, 143–145, 162, 167, 185, 186], two limiting types of relaxation processes can be distinguished: 1) longitudinal relaxation processes in which the anisotropic relaxation elements (electric dipoles, emitting oscillators in PL, and vectors joining the interacting nuclei in NMR) are directed along the backbone of the main chain and 2) transverse processes in which the relaxation elements are normal to the main chain.

In contrast to the simplest idealized models, for actually existing chains with various types of local short-range order (trans-conformers or helical segments) it is impossible to distinguish uniquely the directions "along" the chain and "transverse" to it. Hence, this division is not so rigorous as the division into vibrational symmetry coordinates in the vibrations of regular linear crystals.

However, the division into transverse and longitudinal relaxation processes reflects the anisotropy of structure and local interactions that actually exists in chains and becomes apparent in various phenomena.

The longest times of the longitudinal relaxation spectrum depend on the molecular mass of the chain. In contrast, transverse times for relatively long flexible chains are independent of molecular mass.

Transverse relaxation processes are, to a great extent, related to local conformational changes in the chain. However, local longitudinal processes also exist, e.g. when single "longitudinal" polar groups are included in the chain or markers with longitudinal orientation of the emitting oscillator are present.

At the present state of theory, the discrimination of the actual mechanisms of mobility and the choice of theoretical models are mainly based on the analysis of experimentally observed dependences of characteristic relaxation times on the temperature and viscosity of the solvent and the chemical structure of the chain.

Thus, in the investigation of polarized luminescence of poly(methyl methacrylate) with a luminescent marker in the main chain it has been established that relaxation times τ_w are proportional to the solvent viscosity at a given temperature $T^{[48, 55]}$. It was found that the temperature dependence of τ_w is determined only by the temperature dependence of viscosity $\eta(T)$ (i.e. $\tau_w(T, \eta) \sim \eta(T)$) and does not include any contribution of the barriers to internal rotation. Hence, the conclusion has been drawn[40-42, 48, 170] that the mechanism of motion in the main chain of PMMA is predominantly continuous and "barrierless". This conclusion is valid at least for motions of a segment with a luminescent marker in the time interval of ca. $10^{-9}-10^{-8}$ s.

In contrast to this case, relaxation times, that are to a certain extent related to the motion of a side group with a LM in the same PPMA sample[40-42, 55], are characterized by a temperature dependence differing from that of the solvent viscosity. The effective activation energy $U = d\ln\tau/d(1/kT)$ exceeds the activation energy U_η of the viscous flow of the solvent.

Theoretical conformational calculations of Birshtein, Gotlib and Grigor'eva[46,187-189] are also in favor of the concepts that a great hindrance to rotational isomerization exists in the main PMMA chain and that rotational-isomeric transitions (at $U > 6$ Kcal/mol) in side groups are possible.

These results have led to the suggestion[41, 42, 46, 48, 143, 144] that the motion of the main PMMA chain is actually described by a cooperative continuous mechanism of motion along some equipotential surfaces in the conformational space of the chain. It is the motion of relatively large-chain segments depending on the viscosity of the medium rather than on intrachain barriers.

However, Valeur and coworkers[164] used a lattice rotational-isomeric tetrahedral chain model for the interpretation of the data on *PL* (with the aid of a pulsed method and quenching, in labeled polystyrene.

The validity of this dynamic model can be justified by the fact that in polystyrene and other carbon-chain polymers of the type $(-CH_2-CHR-)_n$(poly(p-chlorostyrene), poly(methyl acrylate) or poly(vinyl acetate)), containing no methyl groups bonded directly to the main chain, rotational isomerization with $U \sim 5-6$ Kcal/mol (in the time interval of $\tau \sim 10^{-8} - 10^{-9}$ s in solvents with $\eta \sim 0.01$ P) can occur[188-192]. This conclusion is also confirmed by theoretical conformational calculations of these polymers.

In PMMA and similar systems such as poly(alkyl acrylates) with long side chains[86], the quasi-continuous "barrierless" character of the motion of the main chain is probably due to the suppression of rotational isomerization owing to the interaction of methyl side groups and side chains and groups. However, other types of macromolecules are possible in which almost free "barrierless" rotation actually occurs depending also on the solvent viscosity. It has been shown[107] that for a labeled polyamidoacid containing a label in the main chain in DMF and in mixed solvents the times τ_w are proportional to η both in the isothermal experiment (in a DMF-water system) and upon variation of the temperature (in DMF).

5.3 Spectra of Relaxation Times and Polarization of Luminescence

As already mentioned in Sects. 1.1 and 1.2, a characteristic feature of intramolecular mobility in polymers is the existence of relaxation time spectra. In this case, the time dependence of the mean value of the Legendre polynomials of the 2nd order $\langle P_2(\cos\theta)\rangle = (3/2)[\langle\cos^2\theta\rangle - (1/3)]$ is given by Eqs. (1.2.9) and (1.2.10).

Just as for other relaxation processes, the continuous distribution function of τ and logarithmic distribution function

$$L_{\langle P_2(\cos\theta)\rangle}(\ln \tau) \equiv L(\tau)$$

are introduced and, hence,

$$\langle P_2(\cos\theta)\rangle = \int L_{P_2}(\tau)\exp(-3\,t/\tau)\,\mathrm{d}\ln\tau \tag{5.3.1}$$

For a comparison with general theory of relaxation properties, it is more convenient to introduce $\tau' = \tau/3$.

The physical reasons for the appearance of relaxation spectra in polymer chains can differ. They comprise: 1) the cooperative character of the motion of a multi-particle polymer chain, 2) the anisotropy of the shape of a kinetic chain segment and consequently the appearance of several relaxation times, 3) the chemical and structural heterogeneity of the chain (copolymers, structurized macromolecules, polymers with side chains, etc., and cross-linked and branched-chain systems).

It follows from Eq. (5.3.1) that the relationship between $1/Y$ and $L(\tau')$ is given by the well-known Steeltjes transform of function $L(\tau')$

$$Y^{-1}(\tau_f) = \int L(\tau')(\tau' + \tau_f)^{-1}\,\mathrm{d}\tau' \tag{5.3.2}$$

The inversion of transform (5.3.2) and the determination of $L(\tau')$, when the analytical dependence $Y(\tau_f)$ is known, have been considered previously[193]. The results were based on the well-known inversion equations of Kirkwood and Fuoss[194] (see also Ref.[195]) establishing a relationship between $L(\tau')$ and the corresponding dynamic compliance function $\hat{Z}(i\omega)$. Indeed, the reduced complex dynamic compliance corresponding to a distribution $L(\tau')$ is given by

$$\hat{Z}(i\omega) = \int L(\tau')(1 + i\omega\tau')^{-1}\,\mathrm{d}\ln\tau' \tag{5.3.3}$$

If the condition of normalization

$$\int L(\tau') \, d\ln \tau' = 1 \tag{5.3.4}$$

is used, Eq. (5.3.2) gives directly

$$1 - Y^{-1} = \int L(\tau') [1 + \tau'/\tau_f]^{-1} \, d\ln \tau' \tag{5.3.5}$$

Comparison of Eqs. (5.3.3) and (5.3.5) leads to a correspondence between $\hat{Z}$ and Y (at given $L(\tau')$)

$$Z(i\omega) = \{1 - 1/Y(\tau_f)\}\big|_{i\omega = \tau_f^{-1}} \tag{5.3.6}$$

Equation (5.3.6) reflects the relationship between Steeltjes and Fourier transforms. Using the Fuoss-Kirkwood equations

$$L(\tau') = \pi^{-1} \operatorname{Im} Z(i\omega = -1/\tau') \tag{5.3.7}$$

we obtain[193] the desired inversion of Eq. (5.3.2)

$$L(\tau') = \pi^{-1} \operatorname{Im} [1 - Y^{-1}(\tau_f^{-1} = i\omega)]\big|_{i\omega = -1/\tau'} \tag{5.3.8}$$

Equation (5.3.8) implies that in order to calculate $Y(\tau')$ it is necessary to represent $Y(\tau_f)$ as a function in a complex plane of a complex variable $\tau_f^{-1} = i\omega$, to replace ω by $-1/\tau'$ and then to find the imaginary part of this expression considering it as a function of $1/\tau'$.

Hence, Eqs. (5.3.5) and (5.3.8) allow both the analysis of the behavior of $Y(\tau_f)$ (or $Y(\tau/\eta)$) at a definite form of the spectrum $L(\tau')$ (and a given dependence of parameters $L(\tau')$ on T/η) and the solution of the reverse problem: the determination of $L(\tau')$ by using the known analytical dependence $Y = Y(\tau_f)$.

If transformations (5.3.7) are used, the complex compliance $Z(i\omega)$ should be given as an analytical function of ω on the whole complex plane. As the theory of irreversible processes shows, $Z(i\omega)$ (and, hence, $Y(\tau_f)$) should exhibit some properties resulting from general principles of dynamics (e.g. the principle of causality) and the Kramers-Kronig reciprocal equations[195].

If $Z(i\omega)$ and $Y(\tau_f)$ are found by a precise solution of a dynamic problem, the analytical properties proper of $Z(i\omega)$ are exhibited automatically. When semi-empirical equations are used for $Z(i\omega)$ of $L(\tau)$ (see below), it should be checked whether the corresponding properties of $Z(i\omega)$ are revealed.

Furthermore, it should be borne in mind that the relationships obtained directly in the form of Eqs. (5.3.7) and (5.3.8) are used in the simplest manner only if the spectrum $L(\tau')$ is really continuous. The case, for which the distribution function of τ contains discrete lines, must be considered specially[201].

The expansion of the function Y into the power series of the parameter τ_f/τ (or τ/τ_f) also suggests the asymptotic behavior of Y at low or high values of τ/τ_f or at low or high values of T/η, depending on the asymptotic behavior of $L(\tau')$ at $\tau \to 0$ or $\tau' \to \infty$ (see Sects. 1.2.14 and 1.2.17).

The relationship between the asymptotic behavior of $L(\tau)$ at τ tends to 0 of τ tends to ∞ and the behavior of $Y(\tau_f/\tau)$ (or $Y(T/\eta)$) can easily be established if one takes into account that the slope of Y in the system of coordinates τ_f or $T\tau_f/\eta$ at $\tau_f/\tau < 1$ is given by the value of $\langle 1/\tau \rangle$ and the slope of the asymptote by the value of $1/\langle \tau \rangle$.

Correspondingly, the slope of $Y = \tilde{Y}/Y_0'$ is given by the value $1/\tau_\omega = \langle \tau \rangle / \langle \tau^2 \rangle$. Various dynamic models of the polymer chain and the corresponding relaxation spectra for the longitudinal component of the emitting oscillator have been considered both in the presence and in the absence of limitations in the thermodynamic and kinetic chain flexibility[39, 93, 156, 170, 171]. It was found that, in a general case, the relaxation spectra are so broad that the values of $\langle \tau^2 \rangle$ and $\langle \tau \rangle$ diverge (for a chain of infinite length). Hence, for a chain of an infinitely high molecular weight the curve $Y(T/\eta)$ exhibits no "theoretical" linear asymptote and at $T/\eta \to \infty$, Y increases more slowly than T/η, e.g. according to $\sqrt{T/\eta}$ (Ref.[164]) or $(T/\eta)/\ln(T/\eta)$ (Ref.[171]).

Only when some assumptions (sometimes relatively arbitrary and not unique) are made about additional torsional-vibrational or other motions superimposed on the segmental micro-Brownian motion of the main chain, is it possible to obtain a faster decrease in $L(\tau)$ at $\tau \to \infty$ or in $\langle P_2(\cos\theta) \rangle$ at $t \to \infty$[156, 164].

In the PL of macromolecules, the same or similar relaxation times and forms of local micro-Brownian motions occur as in local processes determining dielectric, mechanical and nuclear magnetic relaxation.

Hence, it is natural to use in the theory of PL the same "semiempirical" forms of the distribution of τ and of the corresponding values of $Z(i\omega)$ in a complex plane. Some types of distributions $L(\tau)$ and $Y(\tau_f)$ have been obtained and discussed by Gotlib and Torchinskii in Refs.[144, 171, 193]. For instance, the distribution of τ according to Cole-Davidson[196] leads to expressions

$$L(\tau') = \pi^{-1} \sin \pi\gamma \, [\tau_{max}/3\,\tau' - 1]^{-\gamma}$$
$$Y(\tau_f) = 1 + [(1 + \tau_{max}/3\,\tau_f)^\gamma - 1]^{-1} \tag{5.3.9}$$

The inverse (logarithmic scale) distribution of the Cole-Davidson type[144] yields

$$L(\tau') = \pi^{-1} \sin \pi\gamma \, [3\,\tau'/\tau_{min} - 1]^{-\gamma}$$
$$Y(\tau_f) = [1 + 3\,\tau_f/\tau_{min}]^\gamma \tag{5.3.10}$$

As has already been reported[144], the latter type of distribution is obtained by solving a dynamic problem for a model of flexible Gaussian subchains. It is possible to consider also a very general distribution of $L(\tau)$ of Gavriliak-Negami[198].

The Cole-Davidson distribution with the parameter $\gamma \approx 0.4-0.7$ giving the dependence $Y(T/\eta)$ which is the closest to the linear dependence over a wide range of T/η (at $T/\eta > (1-10) \times 10^2 \, (K \cdot cP)$, agrees reasonably well with experimental data on PL. The box distribution (in the $\log \tau$ scale) of Fröhlich's type[199] where $\tau_{min} < \tau < \tau_{max}$ or distributions ranging from $\tau = 0$ to $\tau \to \infty$ but rapidly decreasing at both $\tau = 0$ and $\tau = \infty$ like the Gaussian distribution of Wichert-Eger (see Ref.[195] p. 66)

$$L(\tau) \sim \exp[-\gamma(\ln \tau/\tau_0)^2]$$

may exist, also ensuring the convergence of both $\langle 1/\tau \rangle$ and $\langle \tau \rangle$.

5.4 Shape of Relaxation Spectrum and Evaluation of the Size of Kinetic Units

The theory of relaxation spectra in polarized luminescence for various dynamic models of a flexible polymer chain has been developed by several groups of workers. Wahl[39] has proposed a theory for the model of Gaussian subchains. The authors and coworkers used dynamic chain models consisting of rigid or deformable elements with continuous visco-elastic mechanism of mobility and rotational-isomeric lattice chain models[93, 107, 170, 171].

Dubois-Violette, Monnery, Geny et al. and Taran have developed a theory of *PL* for the rotational-isomeric tetrahedral lattice chain model[156, 157, 164, 165]. A relatively extensive bibliography and a discussion of some papers are included in several reviews[137, 200–202]. Other papers also contain comprehensive references of investigations carried out in recent years[203–206]. The present authors and their coworkers have derived linearized dynamic equations for average projections of chain elements[93, 107, 154, 161, 162, 167, 170–172] and have obtained the time dependences of $\langle P_1(\cos\theta) \rangle = \langle \cos\theta \rangle$ and $\langle P_2(\cos\theta) \rangle = 3/2[\langle \cos^2\theta \rangle - 1/3]$.

For all the models considered, the time dependence of the value $P_2(t)$ is characterized by a spectrum of relaxation times. The greater the relative thermodynamic chain rigidity (i.e. the greater the statistical correlation between the neighboring effectively rigid kinetic units), the broader is this spectrum[46, 167].

It it is assumed that relaxation times in the spectrum are proportional to T/η, the theory predicts a dependence of Y on T/η that can be directly compared with experimental data.

An important property of relaxation spectra of chain models exhibiting limited thermodynamic bending flexibility has been established[144, 171]. It was found that the initial slope of $\langle P_1(\cos\theta) \rangle = \langle \cos\theta \rangle$ and $\langle P_2(\cos\theta) \rangle = 3/2[\langle \cos^2\theta \rangle - 1/3]$ does not depend on the change in the parameter of thermodynamic flexibility. Hence, the initial slope of $Y(T/\eta)$ and the characteristic time $\tau_{init} = \tau_{-1} \equiv \langle 1/\tau \rangle^{-1}$ are determined only by the parameters of a kinetically rigid segment in the continuous model (by its length and effective frictional coefficient) or by the size and microstructure of a set of kinetic units in discrete lattice models. On the other hand, the greater the thermodynamic chain rigidity, the more the plot of $Y(T/\eta)$ deviates from the initial slope.

Experimental data have been analyzed for labeled PMMA with a LM in the main chain in a mixed methyl acetate-triacetine solvent[170, 171]. A comparison of theoretical and experimental curves suggests that the best correlation between theory and experimental data is achieved when it is assumed that the size of the kinetic unit for a viscoelastic chain model is of the same order of magnitude as that of a statistical segment. The statistical segment of PMMA, the size of which is established from the data on light scattering, dynamic birefringence etc., contains ∼6 PMMA units (see [153] p. 287 Russian ed.). Taking into account high hindrance to internal rotation in PMMA[187, 188], this average size of the kinetic unit cannot be considered too high. Since the kinetic unit is comparable in size with the statistical segment, the neighboring kinetic units are also statistically independent and (in terms of a given model) the macromolecule behaves, as regards its relaxation properties, as a "freely jointed" chain of kinetic units.

A similar analysis of the shape of curves $Y(T/\eta)$ for labeled polyamic acid and numberical calculations of the rotatory diffusion coefficients of its units[107] show that the relaxation behavior of a chain is similar to that of a model chain with free internal rotations. This result is in agreement with the data on other conformational and hydrodynamic properties of PAA[108, 207–211] and with theoretical conformational calculations of Birshtein, Zubkov et al.[210, 211].

5.5 Superposition of Relaxation Processes in the Main Chain and in Side Chains of Macromolecules

The simplest method of taking into account the motions of the macromolecule as a whole has previously been applied[40, 93, 117, 171]. It is based on the assumption that the rotation as a whole and internal motions are superimposed independently. This approximation was often used for the description of other relaxation phenomena: dielectric relaxation in polymer solutions, PL in small molecules with internal rotation, in the analysis of hydrodynamic properties[11, 137, 153, 212–214, 227].

A rigorous theoretical approach to the characterization of the molecular-weight dependence of relaxation times in PL has been used only for the simplest models of the polymer chain: for Gaussian subchains without hydrodynamic interactions[93, 171] and by taking into account the effect of hydrodynamic interactions on the PL[216].

The two-time approximation (see Eq. (4.5.1))

$$1/\tau_w = 1/\tau_{whole} (M) + 1/\tau_{int} \tag{5.5.1}$$

is well fulfilled by experiments using aqueous solutions of a labeled non-ionized poly(methacrylic acid), the molecular weight varying from 4.4×10^3 to 3.3×10^5 [93].

For $\tau_{whole} (M)$, the following equation is valid

$$\tau_{whole} = \gamma M \eta [\eta]/RT \tag{5.5.2}$$

where the coefficient γ depends on the selected chain model ($\gamma = 1.2$ for a spherical impermeable particle[153] and $\gamma = 2$ for a Gaussian coil, etc.[217]).

Equation (5.5.1) has also been used successfully and confirmed in Ref.[117] by analyzing the behavior of aqueous solutions of poly(1,2-dimethoxyethylene) in the "coil-globule" transition range upon heating. In this case, it was taken into account that IMM decreases and τ_{int} increases, when the coil becomes more compact during the transition into the globular shape (see Sect. 4.5).

The occurrence of several types of local relaxation processes apparent in the PL of macromolecules with anthracene LG has been established experimentally[40, 48, 55, 57, 218]. The relative contribution to PL of rapid (high-frequency) motions, the duration of

E. V. Anufrieva and Yu. Ya. Gotlib

Table 21. Parameters of the intramolecular mobility for high-frequency ($\tau_{h.f.} \ll \tau_f$) and low-frequency ($\tau \sim \tau_f$) relaxation processes in the main chain and in side chains of PMMA labeled with luminescent markers of different structure

Designation and structure of fragments of main and side chains with LM's	LG$_{I(II)}$ LM$_5$	LG$_{I(\perp)}$ LM$_6$	LM$_{12}$	LG$_{II}$ LM$_2$	LM$_1$	LM$_{10}$
$1/P_0'$	7	8.2	11	9	9	13
$1/P_0$	5.7	5.7	7.3	6	5.7	5.7
$f = \dfrac{1/P_0' - 1/P_0}{1/P_0 + 1/3}$	0.18	0.29	0.33	0.32	0.36	0.55
$\sqrt{\langle\theta^2\rangle}$ (deg)	13	17	19	19	20	30
τ_f (ns)	7.2	7.6	3.4	8.9	4.3	5.8
τ_w [a]	8.3	7.5	5.1	5.5	3.9	31
$\tau_{m.c.}$ (ns)	8.3	7.5	7.5	7.5	7.5	8.3 ÷ 7.5
$\langle\tau_{LG}\rangle$ (ns)			16	21	8	5 ÷ 5.3

[a] The experiments were performed at 25 °C in methyl acetate and in mixed solvents methyl acetate-triacetate of different viscosities; the values of relaxation times are reduced to the value corresponding to "standard" solvent viscosity $\eta = 0.38$ cP

which, $\tau_{h.f.}$, is shorter than the lifetime of luminescence τ_f for the type of LG under consideration, has been determined[58]. Values of the amplitudes of these high-frequency motions $\sqrt{\langle \theta^2 \rangle}$ were established (see Table 21). Fast high-frequency processes with $\tau_{h.f.} < \tau_f$ become apparent in the *PL* of labeled macromolecules in fairly viscous media at low values of T/η, when the slower low-frequency motions do not yet effect the polarization of luminescence (P). The values of $\tau_{h.f.}$ themselves can be determined from the slope of the tangent to curve $1/P(T/\eta)$ at $T/\eta \rightarrow 0$. Only the upper limit of the values of $\tau_{h.f.}$ is usually determined experimentally, because it is not always possible to fulfill strictly the condition $T/\eta = 0$.

However, the assumption that only two main relaxation processes occur in macromolecules: relatively slow motion of the main chain at time $\tau_{m.c.}$ comparable to τ_f, and the above mentioned high-frequency process occurring in the side chains with $\tau_{h.f.} \ll \{\tau_f$ and $\tau_{m.c.}\}$ does not yet make it possible to describe experimental data.

To describe satisfactorily the experimentally observed dependence $Y(T/\eta)$, it is necessary to introduce the assumption[74] that in macromolecules containing luminescent groups in side chains (LG-11) — appart from the main chain motions (at $\tau_{m.c.}$) and high-frequency motions (at $\tau_{h.f.}$) — slow low-frequency motions in the side groups also exist having the characteristic times $\tau_{LG} \sim (\tau_f$ and $\tau_{m.c.})$. This motion is combined with that of the main chain and becomes apparent as a single process with time

$$1/\tau' = 1/\tau_{m.c.} + 1/\tau_{LG} \qquad (5.5.3)$$

At relatively high T/η ($\sim 1-2 \times 10^4$ K/P), when $\tau_w \sim \eta/T$ greatly exceeds $\tau_{h.f.}$, extrapolated to the same T/η a long almost linear part $1/P(T/\eta)$ is observed ($\tau_{h.f.} \sim 1$ ns, when reduced to $T/\eta = 7.8 \times 10^4$ K/P). In this range of changes in T/η, the dependence of $1/P$ on T/η is determined mostly by the contribution of low-frequency motions ($\tau_{m.c.}$ and τ').

The value of $1/P_0'$ obtained by extrapolation of a linear part of $1/P(T/\eta)$ at high T/η to $T/\eta = 0$ determines the contribution f of high-frequency motions displayed already at the beginning of the range of the linear behavior of $Y(T/\eta)$.

In accordance with the above-mentioned model of motion in side chains, a two-time approximation for Y similar to that used in the investigation of the molecular weight dependence was assumed.

The average time τ_w determined experimentally from the slope of linear dependence $Y'(T/\eta)$ is given by Eq. (5.5.3). The data on PMMA listed in Table 21 show how the structure of the side chain bonding the LG to the main PMMA chain and the position of the attachment of this side chain to LG (to the anthracene ring) affect the times characterizing the rate of rearrangements in side chains with LG.

The authors only found a slight anisotropy of rotatory diffusion for emitting oscillators oriented mainly along the chain or normally to it.

A slight difference in the rotatory diffusion times of markers inserted in the chain in different positions with oscillators directed, in the first approximation, along the main chain and normal to it has also been observed by Monnerie and coworkers[204] and Benz and coworkers[169] with the aid of a pulsed method.

Dynamic models of a polymer chain without side groups used so far in the theory of *PL* lead to a larger value of anisotropy of local relaxation times[145, 171, 172, 185, 186]. These theories do not take into account a more isotropic distribution of centers of viscous resistance existing in an elementary kinetic unit (rigid segment) of the real chains. Often, the most bulky groups of monomer units are located on one side of the backbone chain.

Quantitative evaluations show that the transition to chain models with additional centers of viscous friction in side groups (i.e. the introduction of more isotropic elements of the model) leads to a sharp decrease in the anisotropy, at least for thermodynamically flexible chains.

For all the types of side chains considered, with various lengths and flexibilities, the interactions of a side chain with neighboring chain units lead to a gread hindrance to intragroup motions ($\tau_{int.} \gtrsim \tau_{m.c.}$). Simultaneously, a distinct effect of the change in the side chain structure or the transition to a more mobile chain segment on the parameters of IMM is observed both for high-frequency motions and for slow relaxation processes.

Those relationships may be compared with previous data on the correlation between the chemical behavior and the IMM of macromolecules[98, 99, 219]. They indicate possible mechanisms of the influence of the chemical structure of the reaction center and its position in the chain and the types and properties of molecular motions governing the mutual arrangement of the reaction center and of the reagent, on the chemical behavior of the macromolecule and the reactivity of its functional groups.

5.6 Kinetic Volume Effects

The consideration of the influence of kinetic volume effects, i.e. the interactions and contacts between units separated from each other along the chain contour, on IMM is the least developed part of the theory. We think that this problem can be solved most effectively by the methods of numerical experiments which are presently being developed: the computer simulation of molecular motion by the Monte-Carlo method[155, 165, 173, 220–225] or the methods of molecular dynamics[180, 181, 183, 184, 226–231] and Brownian dynamics[168]. It has been shown by Birshtein, Gridnev, Skvortsov and Gotlib using the Monte-Carlo method that the times of local motions become longer, when the energy of the "polymer-polymer" contact is varied (i.e. on passing from a good solvent to a θ-solvent). When the coils undergo volume contraction, not only the average concentration of the units but also their local concentration close to any given unit and the contact strength increase. This effect results in a decrease of mobility.

Recent numerical experiments by the method of molecular dynamics have shown that, for a chain model consisting of particles joined by ideally rigid bonds, the Van der Waals interactions of chain units cause only a little change in the dependence of relaxation times on the wave vector of normal modes of motions, i.e. in the character and shape of the relaxation spectrum. It was found that for the model chain the important relationship

$$\langle P_2(\cos\theta)\rangle = \langle P_1(\cos\theta)\rangle^3$$

remains valid over considerable time intervals[181, 184] both for longitudinal and transversal components of emission oscillators (see Sects. 1.2.3 and 1.2.5).

The molecular dynamic methods can also be very useful in the study of the molecular motions in polymer chains with bulky side groups.

Acknowledgement. The investigations were carried out at the Institute of Macromolecular Compounds of the Academy of Sciences of the USSR (Leningrad) as a result of the work of physical and chemical laboratories by a research team consisting of E. V. Anufrieva, Yu. Ya. Gotlib, M. G. Krakoviak, S. S. Skorokhodov and their colleagues.

The authors are greatly indebted to their colleagues and coworkers who helped in carrying out the synthesis of polymers with luminescent markers, the experiments and the theoretical calculations and in solving the problems discussed in this work.

6 References

1. Vavilov, S. I.: Sobranie sochinenii (Collected papers), Vol. 2, pp. 9, 190, 293, 367. Moscow: Izdatelstvo Akad. Nauk SSSR 1952
2. Levshin, V. L.: Fotolyuminestsentsiya zhidkikh i tverdykh tel (Photoluminescence of liquids and solids). Moscow–Leningrad: Gosudarstvennoe Izdatelstvo Tekhniko-Teoreticheskoi Literatury 1951
3. Feofilov, P. P.: Polyarizovannaya lyuminestsentsiya atomov, molekul i kristallov (Polarized luminescence of atoms, molecules and crystals). Moscow: Gosudarstvennoye Izdatelstvo Fiziko-Matematicheskoi Literatury 1959
4. Stepanov, B. I.: Lyuminestsentsiya slozhnykh molekul (Luminescence of molecules with complex structure). Minsk: Izdatelstvo Akad. Nauk Belorusskoi SSR 1955
5. Stepanov, B. I., Gribkovskii, V. P.: Vvedeniye v Teoriyu Lyuminestsentsii (Introduction in the theory of the luminescence). Minsk: Isdatelstvo Akad. Nauk Belorusskoi SSR 1968
6. Prinsgheim, P.: Fluorescence and phosphorescence. New York: Interscience Publishers 1949
7. Levshin, V. L.: Z. Phys. *26*, 274 (1924); *32*, 307 (1925); Zh. Rossiiskogo Fiz.-Khim.-Obshchestva, part phys. *56*, 235 (1924)
8. Perrin, F.: Compt. Red. *180*, 581 (1925); *182*, 928 (1926); J. Phys. Radium *7*, 390 (1926)
9. Weber, G.: Biochem. J. *51*, 145 (1952)
10. Weber, G.: Adv. Protein Chem. *8*, 415 (1953)
11. Gotlib, Yu. Ya., Wahl, Ph.: J. Chim. phys., phys.-chim. biol. *60*, 849 (1963)
12. Memming, R.: Z. Phys. Chem., Frankfurt *28*, 168 (1961)
13. Jablonski, A.: Z. Phys. *96*, 236 (1935)
14. Jablonski, A.: Acta phys. Polonica *10*, 33, 193 (1950)
15. Bakhshiev, N. G.: Spectroskopiya mezhmolekularnykh vzaimodeistvii (Spectroscopy of the intermolecular interactions). Leningrad: Izdatelstvo Leningradskogo Universiteta 1972
16. Bakhshiev, N. G., Mazurenko, Yu. T., Piterskaya N. V., Izv. Akad. Nauk SSSR, Ser. Fiz. *32*, 1360 (1968)
17. Mazurenko, Yu. T., Bakshiev, N. G.: Opt. Spektroskopiya *28*, 904 (1970); *36*, 491 (1974); *32*, 979 (1972)
18. Samokish, V. A.: Dissertation for the Degree of Candidate of Sciences, Institute of Evolutional Physiology, Academy of Sciences of USSR, Leningrad 1971
19. Anufrieva, E. V., Volkenshtein, M. V., Samokish, V. A.: Dokl. Akad. Nauk SSSR *195*, 1215 (1970)

20. Volkenshtein, M. V.: Molekularnaya Optika (Molecular optics). Moskow-Leningrad: Gosudarstvennoye Izdatelstvo Tekhniko-Teoreticheskoi Literatury, 1951
21. Sarzhevski, A. M., Sevchenko, A. N.: Anizotropiya pogloshcheniya i ispuskaniya sveta molekulami (Anisotropy of light absorption and emission by molecules). Minsk: Izdatelstvo Belorusskogo Gosud. Universiteta 1971
22. Vavilov, S. I., Levshin, V. L.: Z. Phys. *16*, 136, 1923
23. Vavilov, S. I.: Sobranie sochinenii (Collected papers), vol. 1, p. 129. Moscow: Izdatelstvo Akad. Nauk SSSR 1952
24. Tao, T. In: Molecular luminescence. Lim, E. C. (ed.), p. 851. New York, Amsterdam: W. A. Benjamin Inc. 1969
25. Tao, T.: Biopolymers *8*, 609 (1969)
26. Gordon, R. G.: J. Chem. Phys. *45*, 1643 (1966)
27. Nishijima, Y., Asai, T.: Repts Progr. Polym. Phys. Japan *11*, 419 (1968)
28. Onogi, Y., Nishijima, Y.: Repts Progr. Polym. Phys. Japan *14*, 533 (1971)
29. Ward, J. M., 5th Europhys. Conf. Macromol. Phys., Orientation effects in solid polymers, p. 1–26. Budapest, April 27–30, 1970
30. Nobbs, J. H., Bower, D. J., Ward, J. M.: Polymer *17*, 25 (1976)
31. Brestkin, Yu. V., Edilyan, E. S., Frenkel, S. Ya.: Vysokomol. Soedin *A 17*, 451 (1975)
32. Brestkin, Ju. V. et al.: Faserforsch. Textiltechn. *28*, 519 (1977)
33. Fuhrmann, J., Hennecke, M.: Colloid Polym. Sci. *254*, 6 (1976)
34. Terskoi, Ya. A., Zhevandrov, N. D.: Zh. Prikl. Spektroskopii *7*, 88 (1966)
35. Jarry, J. P., Monnery, L.: J. Polym. Sci., Polym. Phys. Ed. *16*, 443 (1978)
36. Weill, G., Hornick, C.: Biopolymers *10*, 2029 (1971)
37. Czekalla, J.: Z. Elektrochem. *64*, 1221 (1960)
38. Weigert, F.: Z. Phys. *5*, 410, 423 (1921); *10*, 349 (1922); Weigert, F., Nakashima, M., Z. Phys. Chem. *34*, 258 (1928)
39. Wahl, Ph.: Thèse, Université de Strasbourg 1962
40. Anufrieva, E. V. et al.: Vysokomol. Soedin., *A 14*, 1430 (1972)
41. Anufrieva, E. V. et al.: Izv. Akad. Nauk SSSR, Ser. Fiz. *34*, 518 (1970)
42. Anufrieva, E. V. et al.: Dokl. Akad. Nauk SSSR *194*, 1108 (1970)
43. Chandrasekhar, S.: Rev. Modern Phys. *15*, 1 (1943)
44. Kramers, A.: Physica *7*, 284 (1940)
45. Gotlib, Yu. Ya., Salikhov, K. M.: Fiz. Tverd. Tela *4*, 1166 (1962)
46. Gotlib, Yu. Ya.: Some relationships of relaxation behaviour of polymer solutions. In: Relaksatsyonnye yavleniya v polimerakh (Relaxation phenomena in polymers). Bartenev, G. M., Zelenev, Yu. V. (eds), pp. 7–24. Leningrad: Khimiya, 1972
47. Gotlib, Yu. Ya., Skvortsov, A. M.: Vysokomol. Soedin. *A 18*, 1971 (1976)
48. Anufrieva, E. V.: Dissertation for the degree of Doctor of Sciences, Institute of High Molecular Compounds, USSR, Academy of Sciences, Leningrad 1973
49. Anufrieva, E. V. et al.: Vysokomol. Soedin., *A 19*, 102 (1977)
50. Kirsh, Yu. E., Pavlova, N. P., Kabanov, V. A.: Europ. Polym. J. *11*, 495 (1975)
51. Kuznetsova, N. P. et al.: Vysokomol. Soedin. *A 19*, 107 (1977)
52. Kivach, L. N.: Dissertation of the Degree of Candidate of Sciences, Belorusski Gosudarstvenii Universitet imeni V. I. Lenina, USSR, Minsk 1975
53. Kozlov, I. N., Sarzhevski, A. M., Khemich: Zh. prikl. Spektrosckopii *9*, 666 (1968)
54. Zhevandrov, N. D.: Trudi FiAN *66*, 123 (1955)
55. Sheveleva, T. V.: Dissertation for the Degree of Candidate of Sciences, Institute of High Molecular Compounds, USSR, Academy of Sciences, Leningrad 1973
56. Anufrieva, E. V. et al.: Dokl. Akad. Nauk SSSR *200*, 358 (1971)
57. Anufrieva, E. V.: Luminestsentnii metod issledovaniya polimerov. In: Spektroskopicheskiye Metodi issledovaniya polimerov. Gul, V. E. (eds.), pp. 35–62. Moskva: Znanie, 1975
58. Anufrieva, E. V. et al.: Vysokomol. Soedin. *A 18*, 2740 (1976)
59. Burstein, L. L., Malinovskaya, V. P.: Vysokomol. Soedin. *A 15*, 73 (1973)
60. Anufrieva, E. V. et al.: Izv. Akad. Nauk SSSR, Ser. Fiz. *36*, 1007 (1972)

61. Clar, E.: Polycyclic Hydrocarbons, New York, London: Academic Press 1964
62. Imoto, M., Nakaya, T.: J. Macromol. Sci. C7, 1 (1972)
63. Krakovyak, M. G. et al.: J. Macromol. Sci. A 12, 789 (1978)
64. Beckwith, L. J., Waters, W.: J. Chem. Soc. 1957, 1001
65. Fahrenhorst, E., Kooyman, E. C.: Rec. Trav. Chim. 81, 816 (1962)
66. Marvel, C. S., Wilson, B. D.: J. Org. Chem. 23, 1479 (1958))
67. Cherkasov, A. S., Voldaikina, K. G.: Spektroskopiya Polymerov, p. 126. Kiev: Naukova Dumka 1968
68. Krakovyak, M. G. et al.: Vysokomol. Soedin. A 17, 1983 (1975)
69. Anufrieva, E. V., Volkenstein, M. V., Koton, M. M.: Zh. Fiz. Khim. 31, 1532 (1957)
70. Lushchik, V. B., Krakovyak, M. G., Skorokhodov, S. S.: Vysokomol. Soedin., A 22, 1904 (1980)
71. Ushiki, H. et al.: Polym. J. 11, 691 (1979)
72. Hawkins, E. G. E.: J. Chem. Soc. 1957, 3858
73. Štolka, M., Yanus, J. F., Pearson, J. M.: Macromolecules 9, 710, 715 (1976)
74. Anufrieva, E. V. et al.: Vysokomol. Soedin. A 19, 2488 (1977)
75. Rembaum, A., Eisenberg, A.: Macromol. Rev. 1, 57 (1966)
76. Krakovyak, M. G. et al.: Europ. Polym. J. 10, 685 (1974)
77. Bunel, C. et al.: Polym. J. 7, 320 (1975)
78. Hamilton, T. D. S.: Photochem. Photobiol. 3, 153 (1964)
79. Hagitani, A. et al.: Ger. Offen, 2 117 058 (1972)
80. Wahl, Ph. et al.: Europ. Polym. J. 6, 585 (1967)
81. Biddle, D., Nordstrom, T.: Ark. Kemi 32, 359 (1970)
82. Parrod, J., Meyer, G.: Compt. Rend. C262, 1244 (1966)
83. Krakovyak, M. G., Ananieva, T. D., Skorokhodov, S. S.: Synth. Commun. 7, 397 (1977)
84. Stolka, M.: Macromolecules 8, 8 (1975)
85. Iwatsuki, S., Inukai, K.: Makromol. Chem. 179, 189 (1978)
86. Anufrieva, E. V. et al.: Vysokomol. Soedin. A 17, 586 (1975)
87. Anufrieva, E. V. et al.: Vysokomol. Soedin. A 19, 755 (1977)
88. Shibaev, V. P. et al.: Dokl. Akad. Nauk SSSR 232, 401 (1977)
89. Skorokhodov, S. S. et al.: J. Polym. Sci., Symposium No. 42, 1583 (1973)
90. Geller, N. M., Nemilova, N. A., Kropachev, V. A.: USSR Auth. Cert., 398 572 (1973)
91. Bawn, C. E. H., Ledwith, A.: Polyalkylidenes. In: Encyclopedia of Polymer Science and Technology. Mark, H. F., Gaylord, N. G., Bikales, N. (eds.), Vol. 10, p. 337. New York: Wiley Intersci. Publ. 1969
92. Cowel, G. W., Ledwith, A.: Quart. Rev. 24, 119 (1970)
93. Anufrieva, E. V. et al.: Vysokomol. Soedin. A 15, 2538 (1973)
94. Klesper, E., Strasilla, D., Regel, W.: Makromol. Chem. 175, 523 (1974)
95. Strasilla, D., Klesper, E.: Makromol. Chem. 175, 535 (1974)
96. Dack, M. R.: J. Chem. Educ. 49, 600 (1972)
97. Krakovyak, M. G. et al.: Vysokomol. Soedin A 22, 352 (1980)
98. Krakovyak, M. G. et al.: Dokl. Akad. Nauk SSSR 224, 873 (1975)
99. Krakovyak, M. G. et al.: Izv. Akad. Nauk SSSR, Ser. Fiz. 39, 2354 (1975)
100. Pautov, V. D.: Dissertation for the Degree of Candidate of Sciences, Institute of Macromolecular Compounds, USSR Academy of Sciences, Leningrad 1978
101. Anufrieva, E. V. et al.: FEBS Letters 55, 46 (1975)
102. Kirmse, W.: Carbene chemistry. New York, London: Academic Press 1964
103. Trozzolo, A. M., Wasserman, E., Yager, W. A.: J. Am. Chem. Soc. 87, 129 (1965)
104. Krakovyak, M. G. et al.: Vysokomol. Soedin. A 18, 1494 (1976)
105. Kozel, S. P. et al.: Vysokomol. Soedin. A 20, 131 (1978)
106. Valeur, B., Rempp, P., Monnerie, L.: C. R. Acad. Sci. C279, 1009 (1974)
107. Anufrieva, E. V. et al.: Vysokomol. Soedin. A 17, 2803 (1975)
108. Zubkov, V. A., Birstein, T. M., Milevskaya, I. S.: Vysokomol. Soedin. A 17, 1955 (1975); A 16, 2438 (1974)
109. North, A. M., Phillips, P. J.: Trans. Faraday Soc. 63, 1537 (1967)
110. Michailov, G. P.: Vysokomol. Soedin. 8, 1377 (1966)

111. Krakovjak, M. G. et al.: Vysokomol. Soedin. B 20, 131 (1978)
112. Geny, F., Koél, C., Monnerie, L.: J. Chim. Phys. 71, 1150 (1974)
113. Nemethy, G.: Angew. Chem. (Intern. Ed.) 6, 195 (1967)
114. Anufrieva, E. V. et al.: Dokl. Akad. Nauk. SSSR 182, 361 (1968)
115. Anufrieva, E. V. et al.: Dokl. Akad. Nauk SSSR 186, 854 (1968)
116. Bychkova, Y. E. et al.: Molekular Biologija 14, 278 (1980)
117. Anufrieva, E. V. et al.: Dokl. Akad. Nauk. SSSR 207, 1379 (1972)
118. Nekrasova, T. N.: Dissertation for the Degree of Candidate of Science, Institute of High
     Molecular Compounds, USSR, Academy of Sciences, Leningrad 1970
119. Anufrieva, E. V. et al.: J. Polym. Sci., Part C, 16, 3519 (1968)
120. Krakovyak, M. G. et al.: Vysokomol. Soedin. A 22, 143 (1980)
121. Anufrieva, E. V. et al.: Vysokomol. Soedin. A 19, 1329 (1977)
122. Anufrieva, E. V. et al.: Vysokomol. Soedin. B 21, 50 (1979)
123. Sveshnikova, E. B., Morina, V. F., Ermolaev, V. L.: Opt. Spektroskopiya 36, 725 (1974)
124. Anufrieva, E. V. et al.: Vysokomol. Soedin. B 19, 915 (1976)
125. Krakovyak, M. G. et al.: Makromol. Chem. (in press)
126. Papisov, I. M. et al.: Dokl. Akad. Nauk. SSSR 199, 1364 (1971)
127. Papisov, I. M. et al.: Dokl. Akad. Nauk. SSSR 208, 397 (1973)
128. Papisov, I. M. et al.: Dokl. Akad. Nauk SSSR 214, 861 (1974)
129. Anufrieva, E. V. et al.: Dokl. Akad. Nauk SSSR 220, 353 (1975)
130. Papisov, I. M.: Dissertation for the Degree of Doctor of Science, Moskovski Gosudarstvennii
     Universitet, Moskva 1975
131. Anufrieva, E. V. et al.: Makromol. Chem. 180, 1843 (1979)
132. Anufrieva, E. V., Glikina, M. V., Sheveleva, T. V.: Vysokomol. Soedin. B 15, 704 (1973)
133. Anufrieva, E. V. et al.: Dokl. Akad. Nauk SSSR 232, 1096 (1977)
134. Morawetz, H.: Macromolecules in solutions. New York: Interscience Publishers 1965
135. Birshtein, T. M., Skvortsov, A. M., Sariban, A. A.: Vysokomol. Soedin. A 18, 1978 (1976)
136. Cerf, R. S.: J. Polym. Sci. 23, 125 (1957); J. Phys. Rad.: 19, 122 (1958); Chem. Phys. Lett.
     24, 317 (1974)
137. Cerf, R.: Adv. Chem. Phys. 33, 73 (1976)
138. Willbourn, A. H.: Trans. Faraday Soc. 54, 717 (1958)
139. Gotlib, Yu. Ya., Salikhov, K. M.: Fiz. Tverd. Tela 4, 2461 (1962)
140. Schatzki, T. F.: J. Polym. Sci. 57, 496 (1962); Amer. Chem. Soc., Polym. Prepr. 6, 6461
     (1965)
141. Pechhold, W., Blasenbray, S., Woerner, S.: Kolloid-Z., Z. Polym. 189, 14, 1963
142. Koppelman, J.L.: Kolloid-Z., Z. Polym. 216–217, 6, 1967; Lunn, A. C., Yannas, I. V.:
     J. Polym. Sci. Phys. 10, 2189 (1972)
143. Gotlib, Yu. Ya., Darinskii, A. A.: Vysokomol. Soedin. A 12, 2263 (1970)
144. Gotlib, Yu. Ya.: Dissertation for the Degree of Doctor of Sciences, Institute of High Molec-
     ular Compounds, USSR Academy of Sciences, Leningrad 1970
145. Darinskii, A. A.: Dissertation for the Degree of Candidate of Sciences, Institute of High
     Molecular Compounds, USSR Academy of Sciences, Leningrad 1974
146. Ferry, J. D.: Viscoelastic properties of polymers, 2nd edit. New York: John Wiley and
     Sons 1970
147. Lauritzen, J. I.: J. Chem. Phys. 28, 118 (1958); Broadhurst, M.: ibid. 33, 221 (1960)
148. Ishida, Y., Jamafudsi, K.: Koll. Z. 177, 97 (1961)
149. Kuhn, W., Kuhn, H., Büchner, P.: Ergeb. exakten Naturwissen. 25, 1, 1951
150. Flory, P. J.: Statistical mechanics of chain molecules. New York: Interscience Publishers
     1969
151. Volkenstein, M. V.: Configurational statistics of polymer chains. New York: Interscience
     Publishers 1963
152. Birstein, T. M., Ptitsyn, O. B.: Conformations of macromolecules. New York: Interscience
     Publishers 1966
153. Tsvetkov, V. N., Eskin, V. E., Frenkel, S. Ya.: Struktura Makromolekul v Rastvorakh
     (Structure of macromolecules in solutions) Moscow: Nauka 1964

154. Gotlib, Yu. Ya., Darinskii, A. A.: Vysokomol. Soedin. *A 11,* 2400 (1969)
155. Taran, Yu. A.: Dokl. Akad. Nauk SSSR *191,* 1330 (1970)
156. Dubois-Violette, E. et al.: J. chim. phys. phys.-chim. biol. *66,* 1865 (1969)
157. Monnerie, L., Geny, F.: J. Polym. Sci., Part C, *30,* 93 (1970)
158. Wahl, P., Meyer, G., Parrod, J.: Compt. Rend. *C 264,* 1641 (1967)
159. Stockmayer, W. H.: Pure Appl. Chem. *15,* 539 (1967)
160. Stockmayer, W. H. et al.: Disc. Faraday Soc. *1970,* N 409, 182; Orwoll, R. A., Stockmayer, W. H.: Adv. Chem. Phys. *15,* 305 (1969)
161. Gotlib, Yu. Ya., Darinskii, A. A.: Fiz. Tverd. Tela *11,* 1717 (1969)
162. Gotlib, Yu. Ya., Darinskii, A. A.: Dinamika diskretnykh povorotno-izomernykh moaelei polimernoi tsepi (Dynamics of the discrete rotameric models of polymer chains). In: Ref.[46], pp. 283–297
163. Iwata, K.: J. Chem. Phys. *54,* 12 (1971)
164. Valeur, B., Jarry, J.-P., Geny, F., Monnerie, L.: J. Polym. Sci., Polym. Phys. Ed. *13,* 667, 675 (1975)
165. Taran, Yu. A.: Vysokomol. Soedin. *A 13,* 2020 (1971)
166. Gotlib, Yu. Ya., Svetlov, Yu. E.: The theory of vibrational and rotational diffusive processes in chains of rotators and polymer chains. In collected articles: Mekhanizmy relaksatsionnykh yavlenii v tverdykh telakh (Mechanisms of relaxation phenomena in solids), Moscow: Nauka 1972 (pp. 215–219)
167. Gotlib, Yu. Ya.: Kineticheskye uravneniya i vremena relaksatsii dlya kontinualnykh modelei polymernoi tsepi postroennoi iz zhestkikh elementov (Kinetic equations and relaxation times for continuous models of polymer chain consisting of rigid elements). In: Ref.[46], pp. 263–282
168. Fixman, M.: J. Chem. Phys. *69,* 1538 (1978)
169. Bentz, J. P. et al.: Europ. Polym. J. *11,* 711 (1975)
170. Anufrieva, E. V., Gotlib, Yu. Ya., Torchinskii, I. A.: Vysokomol. Soedin. *A 17,* 1169 (1975)
171. Torchinskii, I. A.: Dissertation for the Degree of Candidate of Science, Leningradskii Gosudarstvennyi Universitet, Leningrad 1975
172. Torchinskii, I. A., Darinskii, A. A., Gotlib, Yu. Ya.: Vysokomol. Soedin. *A 18,* 413 (1976)
173. Birshtein, T. M. et al.: Vysokomol. Soedin. *A 19,* 1398 (1977)
174. Monnerie, L., Gorin, S.: J. chim. phys. *67,* 400 (1970)
175. Valeur, B. et al.: J. Polym. Sci., Polym. Phys. Ed. *13,* 2251 (1975)
176. Bubnova, L. P., Burshtein, L. L., Shtarkman, B. P.: Vysokomol. Soedin. *A 16,* 2029 (1974)
177. Burshtein, L. L., Malinovskaya, V. P.: Vysokomol. Soedin. *A 20,* 428 (1978)
178. Gotlib, Yu. Ya., Lifshits, M. I., Shevelev, V. A.: Vysokomol. Soedin. *A 17,* 1360 (1975)
179. Gotlib, Yu. Ya., Lifshits, M., Shevelev, V. A.: Vysokomol. Soedin. *A 17,* 1850 (1975)
180. Balabayev, N. K. et al.: Abstracts of short communications of the International Rubber Conference, Kiev, USSR, 1978 (Vol. 1 A, Preprint A 6)
181. Gotlib, Yu. Ya., Darinskii, A. A., Neelov, I. M.: Abstracts of short communications of the International Symposium on Macromolecular Chemistry, Tashkent, Vol. 5, pp. 181–182. Moskow: Nauka 1978
182. Freed, K.: J. Chem. Phys. *64,* 5126 (1976)
183. Balabayev, N. K. et al.: Vysokomol. Soedin. *A 20,* 2194 (1978)
184. Darinskii, A. A. et al.: Vysokomol. Soedin. *A 22,* 123 (1980)
185. Gotlib, Yu. Ya., Darinskii, A. A., Neelov, I. M.: Vysokomol. Soedin. *A 18,* 1528 (1976)
186. Gotlib, Yu. Ya., Darinskii, A. A., Neelov, I. M.: Vysokomol. Soedin. *A 20,* 38 (1978)
187. Grigoreva, F. F., Birshtein, T. M., Gotlib, Yu. Ya.: Vysokomol. Soedin. *A 9,* 580 (1967)
188. Grigoreva, F. P., Gotlib, Yu., Ya., Darinskii, A. A. In: "Sintez, struktura i svoistva polimerov (Synthesis, structure and properties of polymers)". Materials of the XVth Sci. Conf. of the Inst. High Molec. Comp. USSR, Leningrad: Nauka 1970 (p. 160)
189. Grigoreva, F. P.: Dissertation for the Degree of Candidate of Sciences, Institute of High Molecular Compounds, USSR, Academy of Sciences, Leningrad 1971
190. Hässler, H., Bauer, H. J.: Koll. Z. Z. Polym. *230.* 194 (1969); Bauer, H. J., Hässler, H.: Determination of the rotational potential to polystyrene by sound absorption measurements in solutions. Repts. 6th Intern. Congr. Acoustics, Tokyo 1968 (p. 69)

191. Grigoreva, F. P., Gotlib, Yu. Ya.: Vysokomol. Soedin. *A 11,* 962 (1969); *A 10,* 339 (1968)
192. Grigoreva, F. P.: Vysokomol. Soedin. *B 15,* 444 (1973)
193. Gotlib, Yu. Ya., Torchinskii, I. A.: Vysokomol. Soedin. *A 18,* 2780 (1970)
194. Fuoss, P. M., Kirkwood, J. G.: J. Am. Chem. Soc. *63,* 85 (1941)
195. Gross, B.: Mathematical structures of the theories of viscoelasticity. Paris: Hermann 1953
196. Davidson, D. W., Cole, R. H.: J. Chem. Phys. *18,* 1417 (1950)
197. Cole, K. S., Cole, R. H.: J. Phys. Chem. *9,* 34 (1941); Cole, R. H.: J. Phys. Chem. *23,* 493 (1953)
198. Havriliak, S., Negami, S.: J. Polym. Sci, *C 19,* 99 (1966); Polymer *8,* 161 (1967)
199. Fröhlich, H.: Theory of dielectrics. Oxford: Clarendon Press 1958
200. Chen, R. F., Edelhoch, H. (eds.): Biochem. fluorescence concepts, Vol. 1, New York: Marcel Dekker, Inc. 1975
201. Wahl, Ph.: Decay fluorescence anisotropy (in Ref.[200]), p. 1; Wahl, Ph.: Nanosecond pulse fluorometry. In: New Techniques in Biophysics and Cell Biology, Vol. 2. New York: Wiley Intersci. Publ. 1975
202. Rigler, R., Ehrenberg, M.: Quart. Rev. Biophys. *9,* 1 (1976)
203. Rutherford, H., Soutar, L.: J. Polym. Sci., Polym. Phys. Ed. *15,* 2213 (1977)
204. Valeur, B., Monnerie, L.: J. Polym. Sci., Polym. Phys. Ed. *14,* 11, 29 (1976)
205. Nishijima, Y.: J. Macromol. Sci. *8,* 389 (1973)
206. North, A. M.: J. Chem. Soc., Faraday Trans. *1,* 1101 (1972); Chem. Soc. Rev. *1,* 49 (1972)
207. Adrova, N. A. et al.: Poliimidy-novyi klass termostoikikh polimerov (Polyimides – a new class of thermostable polymers) Moscow: Nauka 1968
208. Koton, M. M.: Adv. Macromol. Chem. *2,* 175 (1970)
209. Rudakov, A. P. et al.: Vysokomol. Soedin. *A 12,* 641 (1970)
210. Birshtein, T. M. et al.: Europ. Polym. J. *13,* 375 (1977)
211. Zubkov, V. A., Birshtein, T. R., Milevskaya, I. S.: J. Mol. Struct. *27,* 139 (1975)
212. Stockmayer, W. H.: Pure Appl. Chem. Suppl. Macromol. Chem. *8,* 379 (1973)
213. Cerf, R.: J. Polym. Sci. *12,* 35 (1954)
214. Steiner, R. F., Mac-Alister, A. J.: J. Polym. Sci. *24,* 105 (1957)
215. Cerf, R.: Chem. Phys. Lett. *22,* 613 (1973)
216. Gotlib, Yu. Ya., Svetlov, Yu. E., Torchinskii, I. A.: Vysokomol. Soedin. *A 21,* 1043 (1979)
217. Ptitsyn, O. B., Eizner, Yu. E.: Zh. Techn. Fiz. *29,* 1117 (1959)
218. Nishijima, Y. et al.: J. Polym. Sci. *A 2,* 5, 23, 37 (1967); Teramoto, A., Masayoschi, M., Nishijima, Y.: J. Polym. Sci. *A 1,* 5, 1021 (1967)
219. Anufrieva, E. V. et al.: Vysokomol. Soedin. *B 22,* 129 (1978)
220. Verdier, P. H., Stockmayer, W. H.: J. Chem. Phys. *36,* 227, 1962
221. Verdier, P. H.: J. Chem. Phys. *45,* 2118, 2122 (1966); *59,* 6119 (1973)
222. Krankbuel, D. E., Verdier, P. H.: J. Chem. Phys. *56,* 3145 (1972)
223. Monnerie, L., Geny, F.: J. chim. phys. phys.-chim. biol. *66,* 1691 (1969)
224. Taran, Yu. A.: Dissertation for the Degree of Candidate of Science, Moscow State University, Moscow 1971
225. Stroganov, L. B., Taran, Yu. A.: Vysokomolek. Soedin. *A 16,* 2317 (1974)
226. Balabaev, N. K., Grivtsov, A. G., Shnol, E. E.: Dokl. Akad. Nauk SSSR *220,* 1096 (1975)
227. Shnol, E. E.: Chislennye eksperimenty s dvizhushchimisya moleculami (Numerical experiments with moving molecules). Preprint Inst. Appl. Math. Acad. Sci USSR N 88 (1975)
228. Balabaev, N. K., Grivtsov, A. G., Shnol, E. E.: Chislennye eksperimenty po modelirovaniyu dvizheniya molekul. III. Dvizheniye isolirovannoi polimernoi tsepochki (Numerical experiments on modeling of molecular motion. III. Motion of a isolated polymer chain). Preprint Inst. Appl. Math. Acad. Sci. USSR No. 4 (1972)
229. Rapaport, D. C.: J. Phys. *A 11* (1), No. 8, L 213 (1978)
230. Gunsteren, W. F., Berendsen, H. J. C.: Mol. Phys. *34,* 1311 (1977)
231. Ryckert, J.-P.: J. Comput. Phys. *23,* 327 (1977)

Time-Resolved Fluorescence Techniques in Polymer and Biopolymer Studies

Kenneth P. Ghiggino[1], Anthony J. Roberts[2] and David Phillips[2]

1 Department of Physical Chemistry, University of Melbourne, Parkville, Victoria 3052, Australia
2 Davy Faraday Research Laboratory, The Royal Institution, London W1X4BS, Great Britain

A. Introduction

This review is concerned with the use of the time-resolved behaviour of fluorescence probe molecules in synthetic and biopolymers, and it will be useful to consider brief-ly the nature of fluorescence, and phenomena which can influence decay rates and induce spectral changes.

I. Fluorescence Intensities

Fluorescence is defined simply as the electric dipole transition from an excited elec-tronic state to a lower state, usually the ground state, of the same multiplicity. Mathe-matically, the probability of an electric-dipole induced electronic transition between specific vibronic levels is proportional to R_{if}^2 where R_{if}, the transition moment inte-gral between initial state i and final state f is given by Eq. (1), where ψ_e represents the electronic wavefunction, ψ_n the vibrational wavefunctions, M is the electronic dipole moment operator, and where the Born-Oppenheimer principle of separability of electronic and vibrational wavefunctions has been invoked. The first integral in-volves only the electronic wavefunctions of the system, and the second term, when squared, is the familiar Franck-Condon factor,

$$R_{if} = \int_{-\infty}^{+\infty} \psi_{ef} \, M \, \psi_{ei} \, d\tau \int_{-\infty}^{+\infty} \psi_{nf} \, \psi_{ni} \, d\tau \qquad (1)$$

To a good approximation the electronic integral is zero unless the states i and f are of the same spin, hence electronic absorption in most organic molecules results in the formation of an excited singlet state from the ground singlet state. There are also sym-metry restrictions on transitions between states i and f imposed by the necessity of the electronic integral being totally symmetric if it is not to vanish. For a transition which satisfies this requirement (symmetry allowed), the molar decadic absorption coefficient ϵ will have a maximum value of the order of 10^5 dm^3 cm^{-1} mol^{-1}. The corresponding value for the rate constant for radiative decay, k_R, of the excited state via the spontaneous fluorescence process is given approximately by (2), and more exactly by (3)[180]

$$k_R \ (s^{-1}) \simeq 10^4 \ \epsilon_{max} \ (dm^3 \ cm^{-1} \ mol^{-1}) \qquad (2)$$

$$k_R = 2.88 \times 10^{-9} \ n^2 \ \langle \bar{\nu}_F^{-3} \rangle_{Av}^{-1} \int \epsilon \, d\bar{\nu}/\bar{\nu} \qquad (3)$$

where $\langle \bar{\nu}_F^{-3} \rangle_{Av}^{-1}$ is a measure of the average frequency of the fluorescence, $\int \epsilon \, d\bar{\nu}/\bar{\nu}$ is the area under the absorption curve, and n is the refractive index of the medium in which the experiment is carried out. Use of Eq. (2) for a symmetry-allowed transition gives a value of k_R of 10^9 s^{-1}. In the absence of any other processes de-populating the excited states therefore, the fluorescence decay time of such an ex-cited electronic state would be $1/k_R = 10^{-9}$ s, or 1 ns. This quantity, $1/k_R$ is termed the natural or mean radiative lifetime, τ_R. In practice, because of competing processes (see below), the actual or measured decay time, τ_F, is for complex polyatomic molecules invariably less than the mean radiative lifetime.

For a transition which does not satisfy the symmetry restriction imposed by equation 1, the transition moment integral can be non-zero if a second-order mechanism is invoked which necessitates the excitation of a non-totally symmetric vibration in one or other of the electronic states. This has the effect of reducing the magnitude of the transition moment integral compared with that for a symmetry-allowed transition, and also leads to the absence of the absorption and emission spectral features corresponding to transitions between the vibrationless ground and excited electronic states.

In addition to symmetry restrictions on electronic transitions, there are restrictions caused by the necessity of overlap in space of molecular orbitals for the electron in its initial and final states. Where this overlap is small, for example in $n \rightarrow \pi^*$ transitions in carbonyl compounds, the electronic integral in Eq. (1) is diminished. The "allowedness" of a transition F expressed as an oscillator strength can be summarised as in (4) by a series of factors, f, which relate in turn to spin, (s), overlap (o), p (parity), sy (symmetry), and which have the following approximate values $f_s = 10^{-5}$, $f_o = 10^{-2}$, $f_p = 10^{-1}$, $f_{sy} = 10^{-1} - 10^{-3}$, where F_A is the oscillator strength of a fully allowed transition. For fluorescence then ($f_s = 1$), values of k_R extend from 10^9 s^{-1} for a fully allowed transition, typical say of dyestuffs, through 10^7 s^{-1} for

$$F = f_s \, f_o \, f_p \, f_{sy} \, F_A \tag{4}$$

symmetry forbidden transitions (typical say of aromatic molecules), down to 10^5 s^{-1} and less (for say carbonyl compounds).

The intensity of fluorescence from any excited molecule, although clearly dependent upon the magnitude of k_R as outlined above, is strongly dependent upon internal competing processes. These are shown schematically in the familiar Jablonskii diagram Fig. 1. The unimolecular processes competing with fluorescence are intersystem crossing to the triplet manifold, internal conversion to the ground state, and photochemical reaction. In condensed media vibrational relaxation occurs on a picosecond timescale, and thus only chemical processes with rate constants in excess of 10^{12} s^{-1} will compete with vibrational relaxation. Subsequent to excitation therefore vibrational relaxation is usually complete before electronic relaxation. Internal conversion is usually fast at higher excess energies, but is not of great importance for lower-lying vibrational levels of the first excited singlet state. The principal process competitive with fluorescence is thus intersystem crossing to the triplet manifold of levels. Writing a simple kinetic scheme permits definition of the quantum yield of fluorescence, ϕ_F, and decay time τ_F in terms of first order rate constants defined as in Fig. 1.

Rate Constant

$$M + h\nu \rightarrow {}^1M^* \qquad I_a \tag{5}$$

$${}^1M^* \rightarrow M + h\nu \qquad k_R \tag{6}$$

$${}^1M^* \rightarrow {}^3M^* \qquad k_{ISC} \tag{7}$$

$${}^1M^* \rightarrow M \qquad k_{IC} \tag{8}$$

$${}^1M^* \rightarrow products \; k_D \tag{9}$$

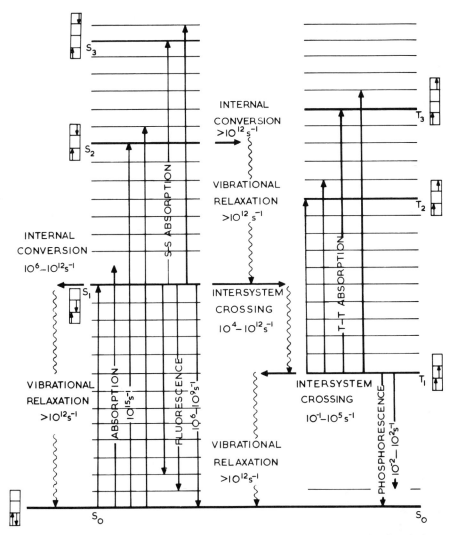

Fig. 1. Jablonskii state diagram illustrating radiative and non-radiative decay paths of excited molecules

From a steady state analysis of the scheme, the quantum yield of fluorescence ϕ_F is given as

$$\phi_F = \frac{k_R}{(k_R + k_{ISC} + k_{IC} + k_D)} \tag{10}$$

The fluorescence decay time, τ_F, is given by (11)

$$\tau_F = (k_R + k_{ISC} + k_{IC} + k_D)^{-1} \tag{11}$$

The intensity of fluorescence seen from any molecule, even before consideration of *bimolecular* interactions, depends upon the magnitude of the rate constant k_R relative to k_{ISC}, k_{IC} and k_D termed Σk. In some molecules, e.g. carbonyls, k_R is very small with respect to Σk and these molecules are thus only weakly fluorescent. In aromatic molecules k_R is usually of the same order of magnitude as Σk, and these are thus strongly fluorescent, and consequently very useful as probes.

II. Fluorescence Spectral Characteristics

The spectral distribution of fluorescence is dictated by the Franck-Condon factors defined as the square of the second term in Eq. 1, and the position of the centre of gravity of the fluorescence depends upon any geometry changes between ground and excited states. The latter point is illustrated in Fig. 2 from which it can be seen that the most probable transition in absorption is to higher energies than that for fluorescence if the potential surface of the excited state undergoes some non-zero displacement with respect to the ground state, and assuming that vibrational relaxation is

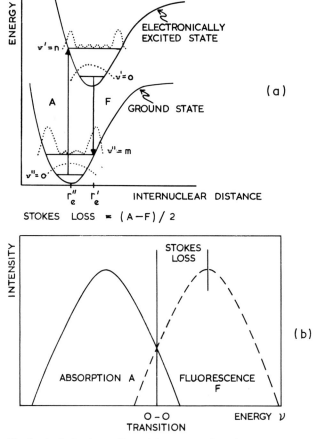

Fig. 2a, b. Stokes loss. a Potential energy surfaces for pseudo diatomic molecules showing most probable transition. b Resulting absorption and emission spectra. (No vibrational structure is shown, although this would be expected in a diatomic molecule)

complete prior to electronic relaxation. If vibrational relaxation is incomplete, then a time-dependent fluorescence spectrum results. However, since in condensed media vibrational relaxation occurs on a picosecond time-scale, time dependent spectra due to this phenomenon can only be observed in picosecond experiments. Solvent relaxation occurs (see below) on a longer time-scale, and can result in the observation of nanosecond time resolved spectra.

For a relaxed system, the difference in energies between centre-of-gravity of the fluorescence and the position of the zero – zero transition is referred to as the Stokes loss, and is clearly a measure of geometry change between relaxed excited and ground states.

Reference to Fig. 1 shows that if vibrational frequencies in ground and excited states are similar, the normal fluorescence spectrum will be a mirror image (on a frequency scale) of the absorption. This condition is not always met. Because of the high density of states in complex polyatmic molecules, resolution of individual vibronic bands in absorption and fluorescence is rarely achieved, spectra being usually broad and relatively featureless. In some molecules of high symmetry where electronic transitions are symmetry forbidden, vibrational structure is observed however.

III. Polarization of Electronic Transitions

The transitional dipole moment operator M contained in Eq. 1 is a vector quantity, or in other words the change in position of the electron in the transition is in a fixed direction with respect to some system of coordinates in which the molecular frame is fixed. Thus the dipole moment operator is resolvable in three directions such that

$$M^2 = M_x^2 + M_y^2 + M_z^2 \qquad (12)$$

and for any transition, only one of these components will be finite. Thus if a perfectly ordered system such as a fixed molecular single crystal is observed in absorption, using plane-polarized light, there will generally be one orientation of the crystal axes with respect to the plane of the polarization of the light which maximizes the absorption probability. Normally fluorescence involves the transition between the same two states as are observed in absorption, and thus the fluorescence is usually polarized parallel to the absorption (i.e. the light emitted has the same directional properties in terms of direction of the electrical vector as has the light absorbed).

In a two-dimensional representation the fluorescence intensity component paralled to the plane of polarization of the electric vector of the incident radiation (the reference direction), $I_{\parallel}$ and that perpendicular to the reference direction $I_{\perp}$ varies with θ, the angle between the reference direction and direction of the optical transition moment (Fig. 3).

It is clearly seen that as θ varies from 0 through $45°$ to approach $90°$, $I_{\parallel}$ varies with I_a (the light absorbed, which varies as $\sin \theta$), and $I_{\perp}$ varies from 0 at $\theta = 0$ through $I_a/2$ at $\theta = 45°$ to approach I_a at $\theta = 90°$. Since I_a, the light absorbed, is also dependent upon $\cos \theta$, it also becomes zero at $\theta = 90\,°C$. The exact magnitude of $I_{\parallel}$ and $I_{\perp}$ as a function of θ can be traced out in a circular diagram shown in Fig. 4.

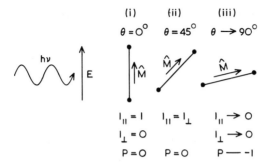

Fig. 3. Polarization of fluorescence in isolated chromophore as function of orientation angle θ of transition moment with respect to plane of polarization of exciting radiation (i) $\theta = 0°$, (ii) $\theta = 45°$, (iii) $\theta = 90°$

It should be noted that this diagram has been constructed for a fixed single molecule, but is clearly also appropriate for a perfectly ordered fixed system with transition moments of all absorbing chromophores parallel. If, as usual, the degree of polarization P is defined as

$$P = \frac{I_{\parallel} - I_{\perp}}{I_{\parallel} + I_{\perp}} \qquad (13)$$

it can be seen that P will vary with θ for this system between the limits of $+1$ and -1 between $\theta = 0$ and $\theta \rightarrow 90°$. For a perfectly random array of absorbing chromophores, it is easily shown that for a parallel (fluorescent) emission, P has the value of $+0.5$ independent of θ, whereas for a perpendicular transition, P = -0.33.

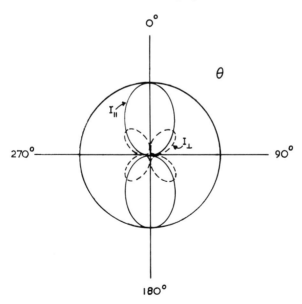

Fig. 4. Traces of $I_{\parallel}$ and $I_{\perp}$ as a function of orientation of perfectly ordered sample with respect of plane of polarization of exciting radiation

Clearly, situations intermediate between perfect order and random distributions occur in arrays of absorbing chromophores, and a treatment is required that allows expression of the orientational distribution of structural units such as crystallites or segments which may be fluorescent in a bulk sample having uniaxial or biaxial symmetry. A complete mathematical approach using a spherical harmonic expansion technique has been developed which expresses the distribution as spherical harmonics of various orders in terms of the Euler angles which specify the orientation of the coordinate system in a fixed structural unit with respect to the coordinate system in the bulk sample[162]. This is of use in solid systems, where time dependence is not observed.

In fluid media, the analysis above is only appropriate if the average rotational relaxation time τ_r is very long compared with the fluorescence decay time, τ_F, ($\tau_r \gg \tau_F$). If the opposite is true, i.e. $\tau_F \gg \tau_r$, all anisotropy introduced in the sample upon excitation is lost through the rotational relaxation prior to emission. If τ_r and τ_F are comparable, then partial relaxation will occur on the same timescale as emission of fluorescence, and thus analysis of the time-dependence of the fluorescence anisotropy can yield information concerning the rotational motion. The emission anisotropy r is defined as

$$r(t) = \frac{I_{\parallel}(t) - I_{\perp}(t)}{I_{\parallel}(t) + 2I_{\perp}(t)} \tag{14}$$

and at time zero, r_0 is given by

$$r_0 = \frac{1}{5}(3 \cos^2 \alpha - 1) \tag{15}$$

where α is the angle between absorption and emission dipoles, which for fluorescence, is usually zero.

In general

$$r(t) = r_0 M_2(t) = r_0 \frac{\langle 3 \cos^2 \theta(t) - 1 \rangle}{2} \tag{16}$$

where $M_2(t)$ is the orientation autocorrelation function, and $\theta(t)$ the angle through which the emission transition moment turns between time zero and t.

For a rigid sphere, Einstein showed that $M_2(t)$ is a single exponential, where D is the rotational diffusion coefficient.

$$r(t) = r_0 e^{-6D_r t}/_\tau \tag{17}$$

The time-averaged value of $M_2(t)$ is

$$\bar{r} = \frac{\int_0^\infty r(t) I(t) \, dt}{\int_0^\infty I(t) \, dt} \tag{18}$$

This leads to the Perrin relationship.

$$\frac{1}{r} = \frac{1}{r_0}\left(1 + \frac{3\,\tau_F}{\rho_{0r}}\right)$$ (19)

or, expressed in terms of degree of polarization

$$\left(\frac{1}{P} - \frac{1}{3}\right) = \left(\frac{1}{P_0} - \frac{1}{3}\right)\left(1 + \frac{3\,\tau_F}{\rho_{0r}}\right)$$ (20)

Thus if the decay time τ_F of a probe molecule is known, then a measure of the single rotational relaxation parameter ρ_{0r} can be made if the time averaged degree of polarization is measured (using continuous excitation), and P_0 is known. In practice, P_0 is not known, and use is made of the relationship $\rho_{0r} = \frac{3\,V\zeta}{kT}$ where ζ is the viscosity of the medium. Measurement of $\overline{P}$ in solvents of different viscosity then permits evaluation of V and P_0.

There are many uncertainties in this procedure, and it seems clear that direct measurement of the time dependence of emission anisotropy through $I_\parallel(t)$ and $I_\perp(t)$ is preferable. Such measurements are described in a later section.

IV. Fluorescence Decay Times

The fluorescence decay time in the absence of any bimolecular interactions which quench the excited state is defined as in Eq. (11), and as has been stated above, is usually less than the mean radiative lifetime τ_R. It is extremely important to recognise that unless a specific single rovibronic level in the excited state is reached upon absorption, (a condition unrealisable except in some gas-phase experiments using extreme narrow bandpass excitation and rotationally and vibrationally cooled molecular species, such as are obtained in a supersonic jet), the decay time observed is that of an ensemble of molecules in different vibronic levels. In condensed media the distribution in the ensemble depends upon the temperature and vibrational energies, each vibrational level possessing its own intrinsic decay time (see for example[175]). Despite the fact that many levels are emitting simultaneously, fluorescence decay times of such ensembles of molecules are invariably well fitted to a single exponentially decaying parameter, τ_F, which is therefore a kinetic parameter descriptive of that ensemble. It is of course possible to synthesize such a single exponential decay parameter by the summation of any number of suitable weighted exponential terms, but such a procedure cannot be performed uniquely and yields no information of physical significance, and is in violation of the generally accepted principle (Occam's razor) that the simplest physical model which is compatible with experimental measurement suffices.

When fluorescence decay curves are obtained which cannot be fitted to a single exponential decay component, but can, after suitable and stringent statistical tests (see Experimental section), be fitted to say, two exponential terms, then it is entirely proper to develop models which are based upon two *kinetically* different species, recognizing that within one kinetic species there may be a whole distribution of molecular species. If as in some circumstances fits to two-component decay are not good, then three components may be necessary, although curve-fitting then becomes an increasingly hazardous procedure. Again, if it is possible *with confidence,* to fit a decay curve to three weighted exponentials it is proper to seek a physical model in terms of three *kinetically* different species.

Fluorescence decay times can often be influenced by bimolecular processes involving ground-state partners M, and impurity molecules Q (see next section), and the general expression in fluid media for a single component decay is thus (21)

$$\tau_F^{-1} = k_R + k_{ISC} + k_{IC} + k_D + k_M [M] + k_Q [Q] \tag{21}$$

Causes of dual (or multi) exponential decay include
a) Excitation of two (or more) non interfering components of different decay times.
b) Interconversion of two (or more) emitting states.
c) Rotational relaxation.

V. Environmental Effects on Fluorescence

The effects of environment upon fluorescence, spectra, yield, and decay characteristics provide the interest in the use of fluorescence probes in macromolecular systems. These effects will be discussed briefly in turn.

1. General Dielectric Effects

The spectral position of absorption and fluorescence are influenced by the dielectric properties of the medium in which observations are made. Figure 5 shows that the vapour phase O–O bands in absorption and fluorescence of a molecule are identical, whereas in solution with solvent of static dielectric constant ϵ, refractive index n, the bands are no longer coincident. The differences can be rationalized as follows. From Onsager theory, a solute molecule of dipole moment μ in a spherical cavity of radius a polarizes the dielectric of the solvent, producing a reaction field. This is given for the ground-state of the solute molecule (of dipole moment μ_0), by (22). Upon excitation, and invoking the Franck-Condon principle, the electronic excitation is much more rapid than the dielectric relaxation time of

$$R_0 = \frac{2\mu_0(\epsilon - 1)}{a^3(2\epsilon + 1)} \tag{22}$$

the solvent, so the reaction field of the Franck-Condon excited state S_1' is given by (23), in which the solvent still experiences the ground-state solute dipole

$$R_1' = \frac{2\mu_0(n^2 - 1)}{a^3(2n^2 + 1)} \tag{23}$$

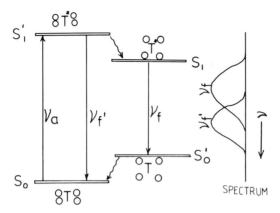

Fig. 5. Schematic of solvent relaxation, with resulting fluorescence spectral shift for solute molecule T

moment, and the high frequency dielectric constant $(= n^2)$ is used in place of ϵ. If the decay time of the excited state of the solute exceeds the dielectric relaxation time, then the equilibrium excited state system of dipole moment μ', has a reaction field given by (24).

$$R_1 = \frac{2\mu'}{a^3} \frac{(\epsilon - 1)}{(2\epsilon + 1)} \tag{24}$$

Following fluorescence, the non-equilibrium ground-state has a reaction field R_0'. Manipulation of these equations shows that the difference in energy of the O–O

$$R_0' = \frac{2\mu'}{a^3} \frac{(n^2 - 1)}{(2n^2 + 1)} \tag{25}$$

transitions in absorption and emission is given by (26). For non-polar solvents $\epsilon \simeq n^2$, and no shifts are observed. For polar solvents shifts can be large. Thus

$$\Delta \bar{\nu} = \frac{2(\mu' - \mu_0)^2}{hca^3} \left[\frac{(\epsilon - 1)}{(2\epsilon + 1)} - \frac{(n^2 - 1)}{(2n^2 + 1)} \right] \tag{26}$$

fluorescence spectral position can be very sensitive to the environment of the fluorescent molecule, and such observations are for example widely used in biological systems to identify different sites in heterogeneous systems such as cell membranes.

Where solvent relaxation is slow compared with the decay time of the fluorescent molecule, time-dependent fluorescence spectra will be observed.

The refractive index of the medium also has an affect upon the radiative rate constant for decay of a fluorophore, as shown in Eq. (27) (see[83]), which can thus affect observed decay times and quantum yields.

$$k_R = \frac{2303 \pi \nu_f^3}{Nc^2} n^2 \int \epsilon \frac{(\nu) d\nu}{\nu} \tag{27}$$

2. Specific Effects

Due to the nature of some excited and ground states, *specific* solvent effects may be observed in some cases. Thus for example in states arising from n → π* excitations, the absorption bands undergo blue shifts in polar solvents relative to those observed in non polar solvents, due to the binding of the non-bonding electrons to the solvent. Where there are two excited states of differing electronic configurations in close proximity, change of solvent polarity can invert the ordering of the excited states, leading to a change in the nature of the fluorescence, which can clearly influence spectral profile, decay time and yield. In mixed solvents, or in heterogeneous systems, level inversion can lead to time-dependent spectra.

3. Concentration Effects

One obvious consequence of increasing the concentration of molecules M in any fluorescence experiment is to increase the probability that energy can migrate from one molecule to another by non-radiative processes, thus influencing the decay

$$^1M^* + M \rightarrow M + {}^1M^* \tag{28}$$

characteristics of the fluorescence. The energy migration may occur through a series of near neighbour random hopping processes, or may occur over long distances through an induced dipole mechanism[50], the probability of which is given by (29), which requires overlap of absorption spectrum of acceptor (here M) and emission

$$k_{ET} = \frac{K}{\tau_R R^6} \int_0^\infty f_D(\bar{\nu}) \, \epsilon_A(\bar{\nu}) \, d\bar{\nu}/\bar{\nu}^4 \tag{29}$$

spectra of donor (here $^1M^*$), and depends inversely on the sixth power of the distance separating the chromophores. τ_R here is the mean radiative lifetime of the donor, and K, a constant given by (30), where κ^2 is an orientation factor arising from the induced dipole nature of the transfer, taking values between 0 and 4, a value of 2/3 being appropriate for a random distribution of molecules.

$$K = \frac{9000 \, \kappa^2 \, c^4 \, \ell n \, 10}{128 \, \pi^5 \, n^4 \, N} \tag{30}$$

Energy migration may reveal itself as a fast component in the decay of fluorescence of $^1M^*$, being often in the picosecond region, but possibly in some polymer systems on the nanosecond time scale. The phenomenon can certainly contribute greatly to the observed depolarization of fluorescence[178].

The interaction of an excited molecule $^1M^*$ with a ground-state may in some instances lead to electronic quenching of the excited state, (31). This concentration

$$^1M^* + M \rightarrow M + M \tag{31}$$

quenching in fluid media leads to the inclusion of the term $k_M[M]$ in the expression for decay time given by (21). In rigid media, where diffusional kinetics are not appropriate, other models are needed (see for example[8]).

In many systems concentration quenching is accompanied by the appearance of a new emission band to the red of the fluorescence of the uncomplexed molecule

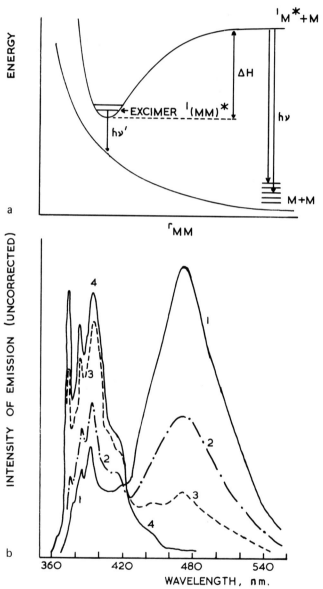

Fig. 6a, b. Excimer fluorescence characteristics (a) potential energy diagram for excimer (b) resulting fluorescence spectra typified by the case of pyrene in fluid solution at concentrations (1) 3×10^{-3} M, (2) 10^{-3} M, (3) 3×10^{-4} M (4) 2×10^{-6} M. [After Parker and Hatchard: Trans. Faraday Soc. 59, 284 (1963)]

^{1}M* which is attributable to the formation of an excited dimer, termed an *excimer,* (32). Excimer fluorescence is characteristically broad, (cf. Fig. 6) and decay characteristics are complex, since excimer formation is reversible

$$^1M^* + M \rightleftharpoons (^1MM^*) \tag{32}$$

Since the modelling of excimer kinetics is very relevant to the discussion of excimer formation in synthetic polymers, detailed examination and analysis of the kinetics of fluorescence decay in these systems is deferred until section C of this review. Suffice it to say here that the time dependence of the fluorescence of the monomer unit ^{1}M* is the *sum* of two exponentially decaying terms, that of the excimer the *difference* between the same two exponentials.

4. Impurity Quenching

In the presence of additive or impurity molecules Q, electronically excited states may be quenched (33), thus in fluid media attenuating the observed fluorescence

$$^1M^* + Q \rightarrow \text{products} \tag{33}$$

decay time of ^{1}M*. Quenching may result from electronic energy transfer, often by the induced dipole mechanism referred to above, from chemical reaction, from enhancement of non-radiative decay brought about by paramagnetic species or molecules with atoms of high nuclear charge, or by complex formation. In the latter case a new fluorescence may be observed which resembles excimer fluorescence in that it is red-shifted and structureless, but is a result charge-transfer to form an excited complex (exciplex). The kinetics of this are exactly comparable to excimer kinetics. In a few cases, weak association between molecule M in the ground-state and an additive molecule may occur. Excitation of the associated complex MQ then

$$M + Q \rightleftharpoons MQ \tag{34}$$

produces an excited state ^{1}(MQ*) which may resemble closely a true exciplex. However, the kinetics of fluorescence in such a system are in principle distinguishable, since in the case of the ground-state complex excitation is instantaneous, and the complex fluorescence does not then exhibit a "growing in" period.

Since molecular oxygen is an ubiquitous impurity, the processes by which it may quench excited states have been widely studied. Quenching of fluorescent singlet states may occur through (35) or (36), with very high efficiencies.

$$^1M^* + O_2(^3\Sigma g) \rightarrow {}^3M^* + O_2(^3\Sigma g) \tag{35}$$

$$\rightarrow {}^3M^* + O_2(^1\Delta g) \tag{36}$$

In a heterogeneous emitting system with two (or more) fluorescence decay components, a "chemical timing" method of achieving time-resolution of fluorescence spectra has been developed which depends upon the selective quenching of the longer-lived molecules by molecular oxygen. This will be discussed in the experimental section below.

VI. Time-Resolved Studies Using Fluorescent Probes

The introductory discussion above will have served to illustrate that the spectral and temporal characteristics of fluorescence are sensitive to environmental influences, and this provides the basis for the use of fluorescence as a probe of molecular environment. In summary, the effects observable are listed below.

1. Fluorescence decay	The number of components observed in decay, with possible causes, are given.
a) Single exponential	single emitting species, decay time sensitive to non-complexing quenchers.
b) Dual exponential	two non-interfering species, reversible complex formation, interconverting states or rotational relaxation.
c) Multiple exponential	Multiple sites, rotational relaxation of asymmetric molecule, combination of several of these.

2. Time-resolved spectra
   Fluorescence spectra will be expected to change with time after excitation for the following reasons.
   a) Vibrational relaxation (a picosecond phenomenon in condensed media).
   b) Solvent relaxation.
   c) Interconversion of electronic states.
   d) Electronic energy transfer to non-identical species.
   e) Complex formation (excimer, exciplex).
   f) Formation of product.
      Examples of these phenomena in the study of macromolecules of both synthetic and biological origin will be given in subsequent sections.

B. Experimental Techniques

As reviewed in the introduction, the luminescence from polymers and biopolymers may be described in terms of spectral shape, quantum yield of emission, decay time characteristics and polarization properties. The recent rapid increase in interest in the usefulness of luminescence techniques to study the structure and properties of molecular systems is partly due to the now ready availability of reliable instrumentation. Although the apparatus necessary for studying the spectral characteristics of luminescence is well established and has been reviewed in detail by several authors[72,104,138], there have been recent rapid developments in the techniques available for time-resolved measurements. These developments have largely been due to the introduction of pulsed lasers as an excitation source to enable reliable observation of luminescence decays in a time domain previously inaccessible.

I. Steady State Luminescence Instrumentation

The basic details of a spectrometer for observing fluorescence and phosphorescence emission are given in Fig. 7. The sample is excited by light of the appropriate wavelength selected from a high intensity light source/excitation monochromator combination. Fluorescence, normally observed at $90°$ to the excitation beam to minimize scatter, is spectrally dispersed through a second monochromator and detected by a photomultiplier tube. The output of the photomultiplier tube is amplified and the signal fed to an X–Y recorder for display. Emission spectra are recorded by scanning the emission monochromator using a fixed excitation wavelength or, alternatively, by scanning the excitation monochromator at a fixed observation wavelength, an excitation spectrum may be obtained.

In order to obtain "true" emission and excitation spectra it is usually necessary to apply corrections for variations in excitation intensity and the wavelength sensitivity of the detection system. The correction needed may be calculated by comparing the instrument response for a standard compound of known corrected spectral characteristics with that of the sample under study, although spectrofluorimeters have been described which fully electronically compensate for intensity and wavelength response of the system[23,139]. Comparison of the area under the corrected emission spectrum with that of various standard fluorescence compounds allows the quantum yield of the luminescence process to be calculated[32,138].

The incorporation of polarizing filters in the excitation and emission paths permits investigation of the polarization properties of the luminescence emission. Such investigations have been found to be of particular use in polymer and biopolymer

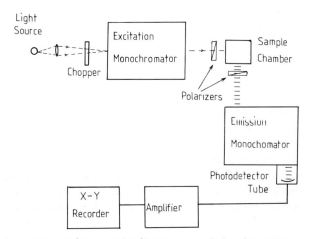

Fig. 7. Block diagram of a typical instrument for measuring fluorescence and phosphorescence emission

systems for assignment of the excited states responsible for the emission process, studies of energy transfer and rotational relaxation behaviour[8, 132, 228]. From an experimental viewpoint it is worth emphasising that the anisotropic nature of elements in the optical path (e.g. lenses, gratings, cells) can lead to serious errors in polarized luminescence measurements unless suitable corrections are made; these will be considered in detail later in this review.

For solution phase studies the sample is usually contained in a quartz cuvette but, for most work, some form of temperature control is desirable to obtain detailed information about energy relaxation pathways. The use of a commercial cryostat to perform luminescence measurements on polymer samples in the range 77° to 373 °K has recently been described by Beavan and Phillips[7] and other home built thermostatted sample chambers have been reported in the literature (e.g.[13,28]).

II. Time-Resolved Techniques

Some degree of temporal resolution of emission may be obtained by incorporating a phosphoroscope attachment in the simple apparatus described above. A mechanical or electronic device is used to allow periodic and out-of-phase excitation and detection of luminescence. In the simplest case a mechanical shutter interrupts the excitation beam periodically and the detection system is gated so that emission is observed only after a fixed interval of time has elapsed after excitation. Under these conditions short-lived processes such as prompt fluorescence will have decayed to zero intensity and only longer-lived emission will be recorded. For mechanical devices the limit of measurable lifetime is of the order of 1 ms, thus allowing time resolved studies to be made of certain phosphorescence and delayed emission processes (see[114, 138]).

Due to the short lifetime of the fluorescence decay process ($10^{-12} - 10^{-7}$s) time-resolved fluorescence studies provide a number of experimental difficulties and require the use of more sophisticated apparatus. Several techniques have been applied to measure fluorescence decay characteristics but all require the use of a pulsed or modulated excitation source (see reviews by Birks[14], Ware[211], Knight and Selinger[109]).

1. Pulsed Sampling Methods

Meserve[128], North and Soutar[135] use most commonly a gaseous discharge lamp to provide light pulses of a few nanoseconds duration to excite the sample and the fluorescence decay is detected by a photomultiplier tube for direct display on a fast sampling oscilloscope. Although simple, the technique is rather insensitive and due to fluctuations in the lamp characteristics and the significant response time of the detection system, lifetimes below 10 ns are difficult to measure in practice. A more recent and elegant extension of this technique is the use of mode-locked picosecond lasers with streak camera detection to study fluorescence processes occurring in the picosecond (10^{-12} s) time domain[43,160,161]. In these systems various harmonics of a single pulse selected from the mode-locked pulse train of a Nd^{3+}/glass laser are used to excite the sample and an ultrafast streak camera detects the luminescence decay. The output display of the streak camera may be optically coupled to an optical multichannel analyser which may be directly interfaced to a computer for rapid analysis of the emission decay kinetics. Although very high temporal resolution may be achieved, the apparatus is costly and measurements are still far from routine.

2. Phase Modulation Techniques

Muller et al.[130], Spencer and Weber[177] and Phillips[148] rely on the detection of the phase lag (δ) between a high frequency (10–50 MHz) modulated excitation light source and the resulting modulation of the emitted fluorescence. For an emission decaying exponentially with lifetime τ, the phase lag is related to the modulation frequency of the exciting light (f) by

$$\tan \delta = 2\pi f \, \tau \tag{37}$$

Because of the high frequency selection available with electronic techniques the accuracy of the method is claimed to be superior than the conventional discharge lamp/pulse sampling systems and a short lifetime limit of 10 ps has been reported[177] However since lifetimes are obtained indirectly by measuring the phase difference, there are some difficulties in analysing multi or non-exponential decays. These difficulties can be overcome if measurement is made of phase and depth of modulation at different modulation frequencies and Weber[218] has recently given an exact solution of the problem of determination of proportion and decay time of N independent non-interacting fluorophores. Experimental precision may not at present be sufficient to permit successful resolution of more than two components but this represents a valuable step forward which may lead to the more widespread adoption of modulation techniques than has hitherto been the case, pulse fluorometry being preferred by most workers at present. This development is of sufficient importance to warrant paraphrasing of the treatment of Weber[218].

The expression for addition of a number of sinusoidally modulated components of the same frequency, but differing in amplitude and phase results in a sinusoidal wave of equal frequency, but with phase angle Φ and square amplitude M^2 related to component amplitudes ϵ_i and phases ϕ_i by (38) and (39).

$$\tan \Phi = \sum_i \epsilon_i \sin \phi_i / \sum_i \epsilon_i \cos \phi_i \tag{38}$$

$$M^2 = (\sum_i \epsilon_i \sin \phi_i)^2 + (\sum_i \epsilon_i \cos \phi_i)^2 \tag{39}$$

In phase fluorimetry, the components are fluorescence from independently exponentially decaying species, and

$$\epsilon_i = f_i \cos \phi_i \tag{40}$$

where f_i is the fraction of total intensity detected.
Thus

$$\tan \Phi = \sum_i f_i \cos \phi_i \sin \phi_i / \sum_i f_i \cos^2 \phi_i \tag{41}$$

and $M^2 = (\sum_i f_i \cos \phi_i \sin \phi_i)^2 + (\sum_i f_i \cos^2 \phi_i)^2$ (42)

For a fixed circular modulation frequency w_r, measurement of the phase difference between exciting light and total fluorescence gives an apparent decay time by phase τ_r^P of

$$\tau_r^P = \tan \phi_r / r$$ (43)

whereas a measurement of depth of modulation yields an apparent lifetime by modulation of τ_r^M given by

$$\tau_r^M = (M_r^{-2} - 1)^{1/2} / w_r$$ (44)

Equations (43) and (44) combined with (41) and (42) permit evaluation of the terms G_r and S_r defined below, corresponding to a given frequency w_r, as (47) and (48) where M_r and ϕ_r' are measured modulation and phase angles at frequency w_r.

$$G = \sum_i f_i \cos^2 \phi_i$$ (45)

$$S = \sum_i f_i \cos \phi_i \sin \phi_i$$ (46)

$$G_r = M_r \cos \phi_r = 1 + (W_r \tau_r^P)^2) (1 + (W_r \tau_r^M)^2)^{-1/2}$$ (47)

$$S_r = M_r \sin \phi_r = G_r W_r \tau_r^P$$ (48)

What Weber has achieved is to show that for N independent components with lifetimes $\tau_1 \ldots \tau_N$ contributing fractions f_1 to f_N of the total intensity, modulation at N different frequencies gives values of G_r, S_r which permit evaluation of individual r's and f's from measurement of M's and ϕ's. Algebraic solutions are given for two and three component systems. Recently a modulated CW laser has been used as an excitation source to achieve picosecond resolution of fluorescence lifetimes using this technique[127].

A related method has been used to demonstrate that lifetimes as short as 200 ps can be measured using the mode noise in a free-running argon-ion laser to produce variations in the excited state population of a fluorophore[78]. Measurement of the rf power spectrum of the resulting fluctuations then reveals the excited state lifetime. Mode noise contains very high frequency fluctuations which the excited state population cannot follow because of its finite lifetime, and thus these high frequency components are absent from the rf spectrum of the fluorescence fluctuations. The fluorescence process thus acts like a low pass exponential filter, and comparison of the fluorescence power spectrum with that of the source provides the decay time data, as demonstrated below.

For a fluorophore, the intensity decaying exponentially as in (49) has a spectral power density in the frequency domain

$$I(t) = I_0 \exp(-t/\tau)$$ (49)

found by Fourier transformation to be

$$P_F(\omega) = \tau^2/(1 + \tau^2 \omega^2) \qquad\qquad (50)$$

The frequency-domain signal is measured in this case at the discrete frequencies at which laser mode beats occur. The time domain representation of these beat frequencies is

$$A(t) = \sum_n A_n \cos(\omega_n t) \qquad\qquad (51)$$

where A_n is the amplitude of the beat at frequency ω_n. Fourier transformation of this yields

$$P_s(\omega) = \sum_n A_n^2 \delta(\omega - 2\pi f_0 n) \qquad\qquad (52)$$

The observed noise spectrum of a fluorescence signal excited by a laser with the beat frequency pattern of Eq. 52 is then the product of (52) and (50), or

$$P_F'(\omega) = (\sum A_n^2 \tau^2 \delta(\omega - 2\pi f_0 n))(1 + \omega^2 \tau^2)^{-1} \qquad\qquad (53)$$

The time-averaged beat frequency of a laser is stable over long periods of time, and thus it is possible to obtain a spectrum (from scatter) proportional to Eq. (52). (Fig. 8a).

Division of the observed fluorescence fluctuation spectrum (Fig. 8b) by such a pattern yields the frequency domain lifetime directly through (50). This is shown for Rhodamine B in Fig. 9 to yield a decay time of 3.2 ns.

There are limits on this method which are imposed by the large spacing of laser mode-noise peaks (130 MKz), setting an upper decay time limit of 7 ns, by detector noise background setting a lower limit of 200 ps, of cost (of a spectrum analyser) and of difficulties in analysis of multi-component fluorescence. Since however any source could in principle be used to measure a noise spectrum, the method is worth further consideration.

3. Time-Correlated Single Photon Counting Methods

i. Equipment[13, 109, 211]

This technique is now very widely used and has found particular application in our own laboratories[57–61,73]. In most arrangements light from a gas filled flash lamp/monochromator or filter combination is used to excite the sample and a fast linear focussed photomultiplier tube detects single photons of emitted fluorescence. The excitation pulse initiates a voltage ramp in a time-to-amplitude converter (TAC), the ramp terminating after a fixed period of time or when a single emitted photon is observed by the photomultiplier tube. The time delay between the excitation flash and the fluorescence event is thus represented by a certain ramp voltage which may be stored as one count in a multi-channel analyser operating in the pulse-height analysis

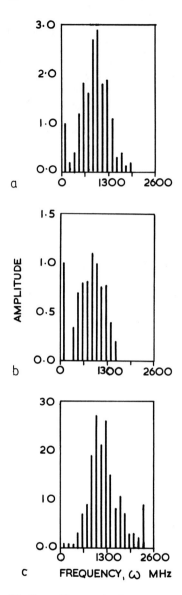

Fig. 8a–c. Time-resolved power spectrum for (a) fluctuations in scattered radiation (b) for fluctuations in fluorescence of Rhodamine B, (c) Rose Bengal; all normalised to the 130 MHz beat. (after Fig. 1, Hieftje et al.)

mode. If the procedure is repeated many times a decay curve will accumulate in the multi-channel analyser representing the probability of emission as a function of time after excitation. In order to obtain a high signal-to-noise ratio and a significant statistical distribution of the data it is desirable to accumulate an absolute minimum of 10^4 counts at the maximum of the fluorescence decay curve, and preferably more. The problems of data analysis are discussed below.

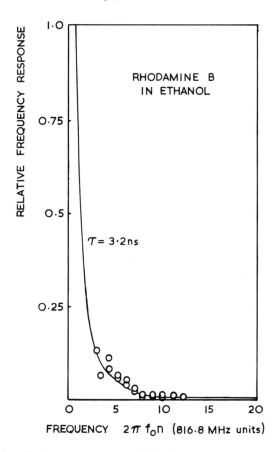

Fig. 9. Frequency response of fluorophore Rhodamine B normalised at first beat frequency. (after Fig. 2, Hieftje et al.)

For the analysis of short lifetimes (less than 10 ns) it is necessary to apply established computational procedures to obtain the true fluorescence decay characteristics[109,150,229]. Although very sensitive a distinct limitation of conventional arrangements is the low intensity, variable pulse profile and low repetition rate of pulse discharge lamps. These factors dictate that, in general, little or no spectral resolution of fluorescence can be obtained, long accumulation times are necessary and time resolution below a nanosecond is rarely achieved. Recently several laboratories have developed instruments based on photon counting detection methods but incorporating actively mode-locked gas lasers and dye lasers as the excitation source[47,57,58,110,176]. These instruments offer a number of advantages over existing time-resolved fluorescence methods. The essential features of the instrument developed in our own laboratories and which has been used for studies on natural and synthetic polymers are shown in Fig. 10.

The excitation source is based on a mode-locked, cavity dumped argon-ion laser (Spectra Physics Model 166/366) and provides a train of narrow, highly stable intense light pulses with repetition rates selectable from single shot to 100 MHz. The wave-

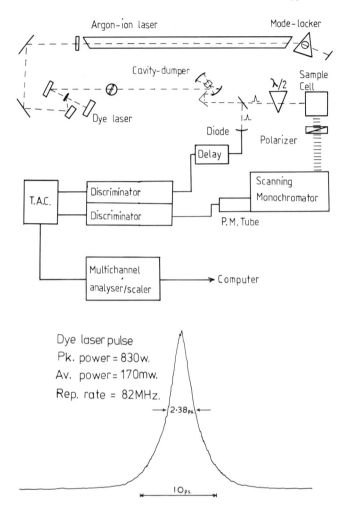

Fig. 10.a Block diagram of a laser excited time-resolved spectrofluorimeter (Ghiggino et al.[57, 59−61]; **b** Background free autocorrelation trace of the light pulse profile from a Spectra-Physics synchronously pumped dye laser system

lengths of light available from the argon-ion laser are restricted to certain argon-ion laser transitions but by synchronously pumping and cavity dumping a dye laser[58,176) a tunable source of picosecond light pulses may be obtained. The characteristics of such light pulses obtained from a Spectra Physics synchronously pumped dye laser system are given in Fig. 10b. In our present system the argon-ion laser transitions at 514.5 nm and 488 nm and Rhodamine dye with emission in the 550−650 nm region have been used. Second harmonic generation of UV pulses is achieved with tempera-ture and angle tuned ADA and ADP crystals to extend the excitation frequencies to 257.25 nm and tunable radiation in the 280 to 320 nm region. The cavity dumping method of pulse selection has certain advantages over other methods in that either narrow sub-nanosecond pulses can be obtained when operated in the mode-locked and dumped mode, or wider but highly reproducible 7 ns pulses may be obtained using the cavity dumped output alone.

Following laser excitation, fluorescence from the sample is spectrally dispersed through a high resolution grating monochromator (Monospek 1000) and single photon counting detection may be used with a TAC/MCA combination to record fluorescence lifetimes. Decay curves may be accumulated with acquisition times orders of magnitude faster than conventional methods due to the high repetition rate (typically 5 MHz) and intensity of the excitation source. A further feature of the laser source is the exact reproducibility of pulse profile throughout the experiment allowing accurate deconvolution of the fluorescence decay kinetics. This is demonstrated in Table I listing the lifetimes of a number of compounds recorded on the apparatus using both narrow mode-locked cavity dumped pulse (FWHM−300 ps) and wider cavity dumped pulses only (FWHM ~7 ns) for excitation. It will be noted that following deconvolution, using a non-linear least squares fitting of the data to a single exponential decay, the lifetimes are identical within the quoted errors using either pulse profile. Fluorescence lifetimes may be reliably measured to 100 ps using these techniques. The inherent stability of the system also allows multi-exponential decays to be analysed with confidence[157−159]. (See below).

ii. Data Analysis

Due to the finite width of the lamp pulse (2−5 ns) and the time jitter in the detection system (voltage discriminators, TAC, photomultiplier tube) the experimental decay $F(t_i)$ is a convolution of the instrument response function and the true decay curve

Table 1. Lifetimes of compounds in degassed room temperature solutions

Compound/Solvent	$\lambda_{FR}{}^a$ (nm)	τ^b (ns)	τ^c (ns)
Acridine/Ethanol	440	−	0.33 ± 0.03
Rose Bengal/Ethanol	600	0.71 ± 0.02^d	−
Anthracene/Cyclohexane	440	4.76 ± 0.05	4.79 ± 0.05
Polystyrene/Dichbromethane	280^e	0.74 ± 0.05	0.76 ± 0.05
	340^f	13.4 ± 0.3	13.9 ± 0.2
9-Cyanoanthracene/Cyclohexane	440	12.7 ± 0.1	12.8 ± 0.1
N-acetyltryptophan methyl amide/water	330	2.95 ± 0.03	3.00 ± 0.03
Rhodamine B/Ethanol	620	3.11 ± 0.03^d	−

a Emission wavelength monitored
b Using cavity-dumped pulses from second harmonic (257.25 nm) of the ion laser only, FWHM ~7 ns
c Using mode-locked cavity-dumped pulses from ion laser, second harmonic (267.25 nm) FWHM ~ 300 ps
d Using cavity-dumped dye laser pulses (580 nm) FWHM ~7 ns
e Corresponding to monomer emission only
f Corresponding to excimer emission only

$$F(t_i) = \int_0^{t_i} G(t') P(t_i - t') dt' \tag{54}$$

in which F(t) is the measured decay curve

          G(t) is the true decay function

and   P(t) is the measured excitation profile, also known as the instrument response function.

Although deconvolution is a well defined mathematical procedure, its application to fluorescence decay curves is attended with numerous difficulties owing to the counting errors and instrumental distortions that accompany single photon counting data. It is now generally accepted that least squares iterative reconvolution is the most satisfactory method of analysing nanosecond decay data[108, 136, 229]. In its simplest form, deconvolution involves solution of Eq. (54) above.

In the least squares fitting procedure, G(t), assumed to be of specific functional form with adjustable parameters, is convoluted according to Eq. (54) with the measured function, P(t). The calculated curve C(t) is compared with the measured curve F(t) and the reduced chi-square, χ_ν^2, is calculated.

$$\chi_\nu^2 = \frac{\sum_{i=1}^{n} W_i^2 F(t_i) C(t_i)^2}{n - p} = \frac{\sum_{i=1}^{n} (r_i)^2}{n - p} \tag{55}$$

W_i is the weighting factor for the ith data point $\left(= \frac{1}{\sqrt{F(t_i)}}\right)$, n is the number of channels over which fitting is performed and p is the number of adjustable parameters in the fitting function. The values of the adjustable parameters are varied until a minimum value of χ_ν^2 is obtained. If the minimum value of χ_ν^2 is close to 1 the function assumed for G(t) is taken to be correct and the final values of the adjustable parameters are assumed to have real physical significance.

For single exponential decay functions with reasonably long decay times (> 2 ns) deconvolution poses few difficulties. Double exponential decays, for which G(t) contains 4 variable parameters, are much more difficult to analyse correctly and a number of tests have been proposed to ensure that the fitting procedure does not mask distortions in the data or the presence of a further component. The most common of these tests is a plot of the weighted residuals, which should be randomly distributed about zero. Also of use is a plot of the auto-correlation function of the residuals[38].

It is worthwhile stressing the need for great care in analysis of even single exponential decay curves, since criteria used by some authors are certainly insufficient. In heterogeneous systems such as polymers, it would be expected that multiple decays might be encountered frequently.

While the analysis of single and double exponential decays may now be regarded as fairly routine the successful deconvolution of triple exponential functions is fraught with difficulties. Correlation among the 6 variable parameters in G(t) cannot be avoided and unless strict precautions are taken the decay function recovered from deconvolution may have little resemblance to the true fluorescence decay[69]. It is, of course, necessary to have data that are as statistically accurate and as free from instrumental artefacts as possible. A systematic study is at present under way to try to ascertain the number of counts that must be collected in order to establish triple exponential functions to satisfactory accuracy. For the present we can say that, for most of the decays we have studied, 30000 counts in the channel of maximum counts is a lower limit.

Even when data are statistically very accurate and free from instrumental distortions triple exponential analysis may be meaningless. For instance in a trial test with simulated data the function

$$G(t) = \sum_{j=1}^{3} a_j e^{-t/\tau_j} \qquad (56)$$

with $a_1 = 0.3$, $\tau_1 = 1.5$, $a_2 = 0.2$, $\tau_2 = 5.0$, $a_3 = 0.1$, $\tau_3 = 10.0$ was convoluted with a real pump profile of width 7 ns and Gaussian noise was added. Deconvolution using our triple exponential routine failed to recover the correct decay parameters although the χ_ν^2 value was quite satisfactory (1.03). It must therefore be admitted that triple exponential analysis is not possible for certain combinations of decay parameters. However, for decay parameters which are reasonably well separated tests with simulated data demonstrated the ability of the routine to recover the correct parameters as well as showing the inability of a two component function to fit three component data of the type encountered in some polymer systems. As an illustration, Fig. 11a shows a two component fit to the three component decay generated by convolution of a real (laser) pulse profile with a $G(t)$ in which $a_1 = 0.25$, $\tau_1 = 2.5$, $a_2 = 0.07$, $\tau_2 = 10.00$, $a_3 = 0.025$, $\tau_3 = 40.0$. The plot of residuals (Fig. 11b) and autocorrelation of residuals (Fig. 11c) are clearly diagnostic of an inappropriate fitting function. The fit to a three component decay is shown in Fig. 12a with residuals and autocorrelation of residuals plotted in Figs. 12b and 12c. All plots are satisfactory and in fact the parameters recovered with this analysis were $a_1 = 0.25$, $\tau_1 = 2.54$, $a_2 = 0.07$, $\tau_2 = 9.72$, $a_3 = 0.026$, $\tau_3 = 39.44$, in good agreement with the correct values. Similar tests with many other trial functions lead to the conclusion that, with the pumping pulse profiles obtained from the present excitation source and decay functions containing decay constants of the magnitude met in many polymer systems under investigation, triple exponential analysis produces meaningful results if statistical accuracy is high and instrumental artefacts are maintained at a minimum. (A small amount of stray light, for instance, would render the analysis meaningless).

In order to increase the level of confidence, a number of other statistics in addition to χ_ν^2 should be calculated. These include the standard deviation of the weighted residuals, the mean residual, a skewness of fit parameter, and, especially, the Durbin-Watson factor, DB[35, 36]. By using DB, calculated according to (57)

$$DB = \frac{\sum_{i=2}^{n} (r_i - r_{i-1})^2}{\sum_{i=1}^{n} r_i^2} \qquad (57)$$

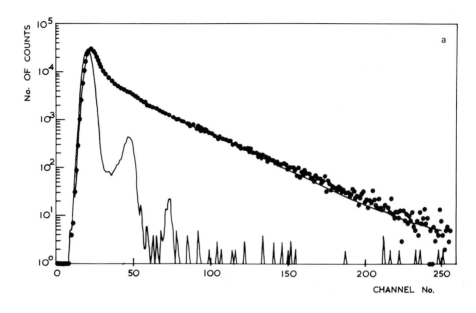

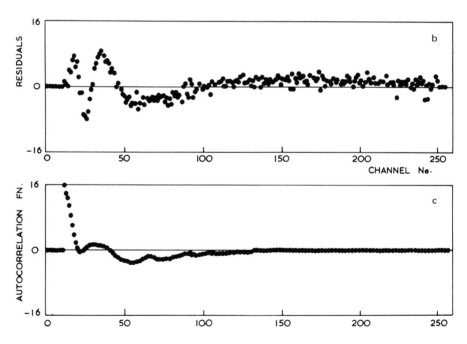

Fig. 11a, b. Two-component fit (**a**), and resulting residuals (**b**) and autocorrelation trace (**c**) to synthesized triple exponential (six-parameter) decay curve (see text)

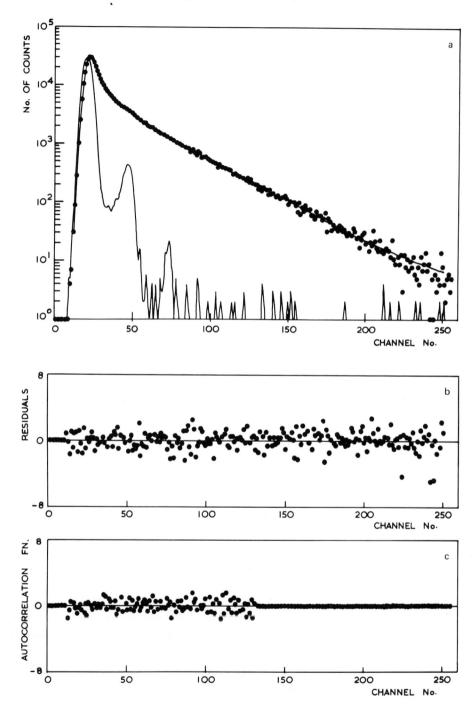

Fig. 12a–c. Three-component fit (a), residuals (b) and autocorrelation trace (c) of same synthetic decay curve as in Fig. 11. Note improvement in fitting, and that realistic values of the six parameters are recovered

and the table of values given by Durbin and Watson[36] one has an additional criterion concerning the suitability of the chosen fitting function.

4. Time-Resolved Fluorescence Spectra

Time-resolved emission spectra (TRES) in the nanosecond time domain are a useful means of studying complex fluorescence spectra. They have been used in the study of monomer-exciplex equilibria[57], ground state complex formation[21,74], solvent relaxation kinetics[38,55,212], acid-base equilibria in the excited state, and overlapping spectra where two or more species are primarily excited[117]. In a single-photon counting instrument such spectra may be obtained directly by setting upper and lower level discriminators on the output pulses of the time-to-amplitude converter and operating the multichannel analyser in the multichannel scaling mode with the channel advance synchronised with the wavelength scan of the emission monochromator. As has been stated[38] this method results in distorted spectra because of the convolution of the excitation pulse with the decay of the sample. In the alternative method[38,151] spectra are not obtained directly but are constructed from deconvoluted decay curves from which the distortions arising from the finite width of the pumping pulse have been removed. With this method, decay curves must be collected at a number of wavelengths (the number depending on the degree of spectral resolution required) spanning the emission spectrum of the sample. For low emission quantum yield samples and conventional flash lamp excitation this procedure can be tedious. Direct measurement of the spectra on the other hand can usually be accomplished fairly rapidly and invariably results in superior spectral resolution. In addition, the spectra of most interest are usually those at very early times, when the excitation pulse approximates a delta function, or at very late times, when the effect of the excitation pulse width may be negligibly small. It is of some importance to ascertain to what extent distortions are present in early time and late time spectra and to determine the usefulness of such spectra in quantitative fluorescence analysis.

A full comparison of the aforementioned methods of obtaining TRES has not hitherto been carried out, presumably because of the time involved in collecting the requisite number of decay curves with conventional single-photon counting instruments. With highly intense, 5 MHz laser pulses used as an excitation source data accumulation can be achieved rapidly and it is worthwhile describing a comparison between the two methods of obtaining TRES.

i. Method I

The output of the time-to-amplitude converter (TAC) in a single photon counting experiment consists of pulses whose intensity is proportional to the time difference between the arrival of a signal synchronised to the rise of the excitation pulse and the arrival of a signal corresponding to a photon emitted from the sample as a result of that exciting pulse. By setting a lower level discriminator at voltage v_1 say, in the multichannel analyser, to which TAC pulses are routed, we exclude analysis of photons emitted at times less than t_1. Similarly we discriminate against photons emitted at times greater than t_2 by setting an upper level discriminator in the analyser at voltage v_2 say. The gate width $t_2 - t_1$ is referred to as δt while the delay time $t_1 + (t_2 - t_1)/2$ is denoted by Δt. By spectrally analysing the photons in the gate width δt we obtain a time-resolved spectrum. If t_1 is very short we refer to the spectrum as an early gated sepctrum, EGS. If t_1 is long the spectrum is a late gated spectrum LGS. The spectral resolution of TRES, collected in this way, referred to as Method I, is dependent in the usual way on monochromator slit width and scanning speed; it may, however, be limited by the channel advance speed in the analyser. The time-resolution is determined by the gate width, δt, and pulse profile.

Because the excitation is not instantaneous, the observed decay curve I(t) obtained with the analyser operating in pulse height analysis mode is a convolution of the excitation pulse profile P(t) with the true decay of the sample G(t) viz.

$$I(t) = \int_0^t P(t') G(t - t') \, dt' \tag{58}$$

When TRES are measured using Method I we in effect take time slices from I(t) at various wavelengths and arrange the sum of photons in those slices as a function of wavelength. We have therefore a distorted spectrum. Similar spectra can be obtained using a gated discriminator circuit[171].

ii. Method II

To obtain a true spectrum, I(t) is actually measured at a number of wavelengths and G(t) obtained by deconvolution[136, 229]. Then the individual G(t)s are scaled to the intensity in the total spectrum and photons in the derived time slice are summed and arranged as a function of wavelength. This method will be referred to as Method II. Spectral resolution is now determined by the number of wavelengths at which I(t) was measured while time resolution is limited only by the channel width in the decay curve measurement.

Deconvolution involves a number of serious difficulties particularly if physically significant decay times are to be assigned to G(t)[17,90,108,150,207]. In the use of Method II some of these difficulties are avoided since only an accurate representation of G(t) is required and no physical significance need be attached to the parameters in a trial function, if such a function is used. But experimental distortions in the decay curve cannot be disregarded. Perhaps the most common difficulty encountered is the wavelength variation of the detection photomultiplier response[150, 207] which is thought to introduce errors mainly in the rising edge of decay curves[207]. Unless corrections are made, any function that is determined using the entire curve may not be a true representation of the sample's decay and may lead to incorrect EGS. In fitting techniques it is common to analyse only over the decaying part of the curve and thus avoid rising edge effects. However, in doing this, information residing in very short times will be lost and EGS may again be incorrect. In the instruments using laser excitation, the pump profile must be collected at the excitation wavelength; hence some distortions are present in the measured I(t)s. Since none of the suggested corrections can be used with confidence TRES obtained even with Method II may have some slight distortions. The procedure in Method II used by us[124] is outlined below.

The decay law $G(t)$ was extracted from the experimental decay curve using non-linear least-squares iterative reconvolution. The parameters were varied until the χ^2_ν values were minimised and initially the trial function for $G(t)$ was chosen to be a sum of 10 exponential terms[213], viz

$$G(t) = \sum_{j=1}^{10} A_j e^{-t/\tau_j} \tag{59}$$

with the τs fixed and the a_js the variable parameters. Quite good fits, as indicated by the value of χ^2_ν were obtained over the fitted region but at long times fluctuations, obviously associated with the deconvolution technique, were observed in the extracted $G(t)$s. Similar fluctuations have been previously reported[213] and are a consequence of the lack of constraint on the fitting function in the region of low counts. Better results could probably have been obtained by choosing a different set of τ_js[213] or by fitting only over the decay of the experimental curve. We chose instead to reduce the number of exponential terms to three[38] and to vary the a_js simultaneously. Although in general χ^2_ν values were somewhat larger than for the ten-exponential function the reconvoluted curves were still satisfactorily close fits to the observed data and the $G(t)$s were well-behaved at all times. It was decided to use this $G(t)$ in the

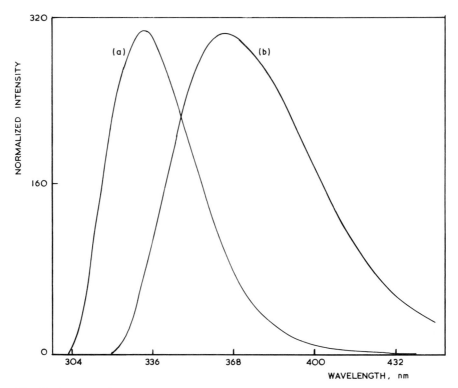

Fig. 13. Normalized fluorescence spectra (a) 5-Methoxyindole in undegassed H_2O; (b) 2-Amino-pyridine in undegassed H_2O

construction of TRES since the fluctuations in the 10-exponential functions at long times put an unacceptably low limit on the Δt at which LGS could be obtained. It should be noted that no physical significance is attached to the three exponentials in the $G(t)$s extracted in this way.

The results of carrying out both method I and II upon a mixture of 5-methoxy-indole (MOI) and 2-amino pyridine (AMP) in water are shown below[124].

In Fig. 13 are shown the spectra, measured from separate solutions of MOI and AMP in water. The lifetimes of MOI and AMP, measured by fitting decay curves from the separate solutions to single exponential functions, were 4.48 ns and 9.57 ns respectively. Therefore the spectrum obtained from a mixture of MOI and AMP in water presented in Fig. 14 should be resolvable into the two spectra of Fig. 13 using Method I. The TRES depicted in Fig. 15 illustrate this separation. It will be seen that the EGS shows a broadening to the red compared to the MOI spectrum while the LGS is broader to the blue than the AMP spectrum. This is expected in view of the fact that AMP makes some contribution to the emission even at extremely short times while at a Δt of 54 ns there is still some fluorescence from MOI. These spectra clearly demonstrate that TRES collected in this way are a good diagnostic tool in examining complex spectra but will not completely separate the individual components when the lifetimes of each species are not very different.

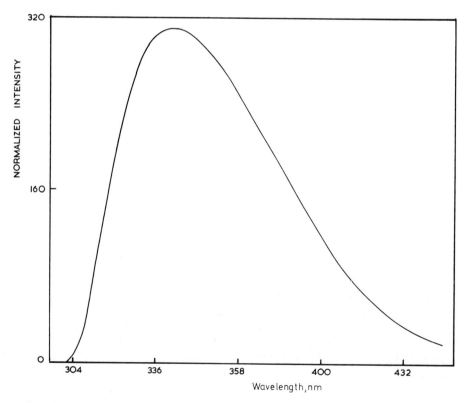

Fig. 14. Total fluorescence spectrum: 5-Methoxyindole + 2-Aminopyridine in undegassed H_2O

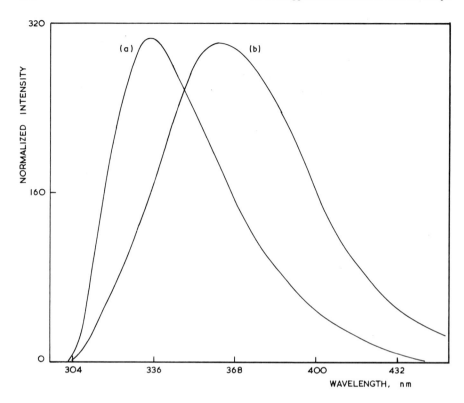

Fig. 15. TRES (Method 1) of MOI + AMP in undegassed H_2O (a) EGS, $\Delta t = 1.28$ ns, $\delta t = 5.76$ ns; (b) LGS, $\Delta t = 53.8$ ns, $\delta t = 7.04$ ns

Decay curves from this mixture were collected at 15 wavelengths from 310 nm to 450 nm. Three examples are shown in Fig. 16. Also shown are the reconvoluted $C(t)$ and the decay law $G(t)$. As is expected, as wavelength increases the intensity at long times in $G(t)$ increases and non-exponentiality becomes more apparent. Areas for the 15 $G(t)$s were calculated according to

$$\text{Area} = \sum_{j=1}^{3} a_j \, \tau_j \tag{60}$$

Each function was divided by its area multiplied by the intensity in the total spectrum (Fig. 14) at the corresponding wavelength. Using these normalised functions TRES were constructed; three examples are shown in Fig. 17. The poor spectral resolution is obvious but is not however, in the case of unstructured spectra such as these a serious disadvantage. If the EGS in Fig. 17 is compared with the MOI spectrum in Fig. 13 it will be seen that the TRES contains some contribution from AMP fluorescence. Δt for this spectrum was 1.28 ns, i.e. 2 channels while δt was 0.64 ns. In order therefore to separate the MOI spectrum completely it would be necessary to collect

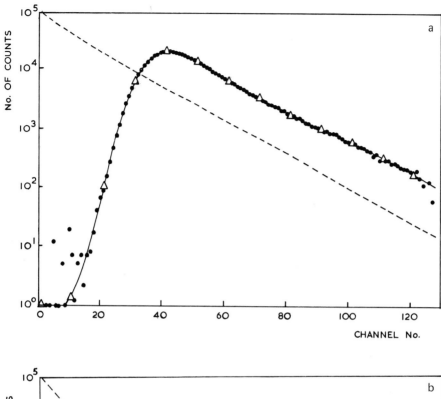

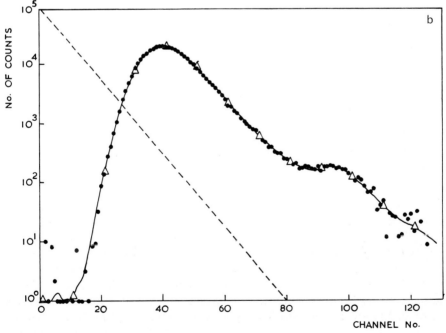

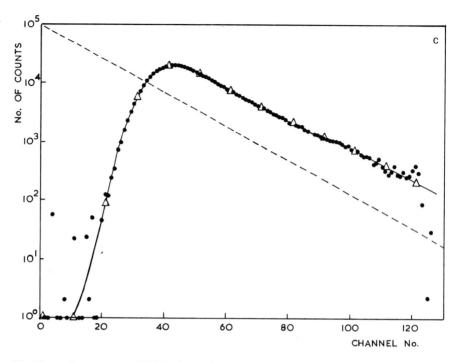

Fig. 16a–c. Decay curves of MOI + AMP in undegassed H_2O —•—•—•— observed decays, 0.64 ns/channel; ——— reconvoluted curves; ------ decay functions, $G(t)$; (a) 310 nm, (b) 380 nm, (c) 450 nm

decay curves on a much smaller time scale. The LGS on the other hand ($\Delta t = 56.3$ ns) is quite a good approximation to the AMP spectrum.

iii. "Chemical" Timing Techniques

Time-resolved fluorescence spectra can be obtained without recourse to sophisticated equipment by the use of an electronic quencher, (invariably molecular oxygen), which will quench preferentially those molecules with the longest decay times, as is illustrated by the Stern-Volmer Eq. (61). Addition of increasing amounts of quencher to a system under continuous illumination thus progressively

$$\frac{(\Phi_F)_0}{\Phi_F} = 1 + k_Q\,\tau[Q] \tag{61}$$

restricts observation of fluorescence to those molecules which have the shortest decay times, thus providing early gated spectra (EGS). The method has been used widely to observe fast processes in solution[105, 115, 164]. A major disadvantage is the inability to observe late-gated spectra, but this is outweighed by the simplicity of equipment required.

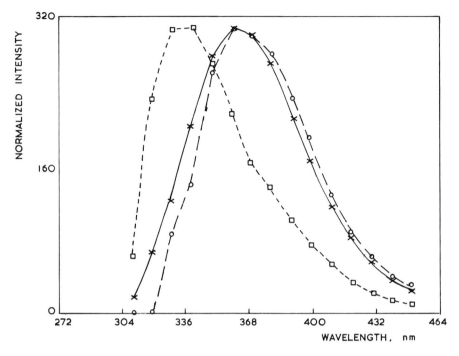

Fig. 17. TRES (Method II) of MOI + AMP in undegassed H_2O TRES constructed from
□ EGS, $\Delta t = 1.3$ ns, $\delta t = 0.64$ ns; x $\Delta t = 19.2$ ns, $\delta t = 0.64$ ns; ○ LGS, $\Delta t = 56.3$ ns, $\delta t = 0.64$ ns

5. Time-Resolved Polarization Measurements

The plane-polarized light pulses characteristic of mode-locked lasers also provide an ideal excitation source for time-dependent fluorescence depolarization studies although conventional excitation sources can be used. If the rotational relaxation time of the excited molecule is comparable to its fluorescence decay time, then the vertical ($I_{\parallel}$) and horizontal ($I_{\perp}$) components of the fluorescence decay observed through suitable polarizers following excitation by polarized light pulses, may be analysed to provide information concerning the size and motion of the molecule[132] and Sect. 5. However, if only the true fluorescence decay characteristics are of interest it is necessary to compensate for these emission anisotropy effects[176, 178]. Perhaps the simplest technique is to analyse only that component of fluorescence emitted at 54.7° to the direction of polarization of the excitation source, the so-called "magic-angle"[178].

Single-photon counting techniques again are of great value in such measurements where intensites are in general low.

The present limitations in time resolution for the time-correlated photon counting technique are due to the time jitter in the detection electronics and the transit time spread in the photomultiplier tube (~ 500 ps). With future improvements in these components and using cw mode-locked lasers as an excitation source, deconvolution of fluorescence lifetimes to a few tens of picoseconds should be achieved. Alternatively the application of repetitively scanning streak cameras or correlation techniques should allow picosecond lifetime measurements to be made[47].

C. Fluorescence from Synthetic Polymers

The classification of Somersall and Guillet (1975) distinguishes between synthetic polymers in which the repeat unit contains an absorbing chromophore, and those in which isolated chromophores are attached to the polymer chain as end-groups or minor components of copolymers, or as impurity sites. Very little time-resolved work has been done on the latter class of polymer, and thus attention is focused here on the first variety. The chromophores which are strongly fluorescent are invariably aromatic in nature, and although there are some polymers with aromatic moieties in the backbone of the polymer chain, such as poly (sulphones) I and poly (ether sulphones) II, very little time-resolved work has been carried out (see for example[1]). The discussion in this section is thus confined exclusively to homo and co-polymers with an aromatic moiety pendant from the polymer backbone. The most important attribute of such polymers is their exhibition of electronic energy migration and their capacity for formation of intramolecular excimers (see introduction[51-54]). Since the kinetics of excimer formation will feature strongly in subsequent discussion, the appropriate scheme for a simple molecular system will be discussed here.

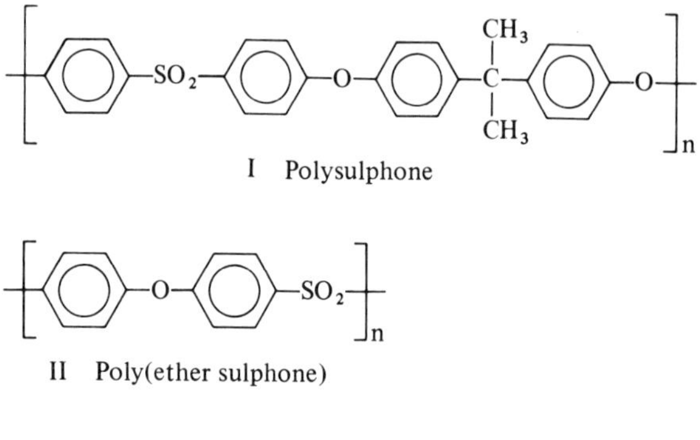

I   Polysulphone

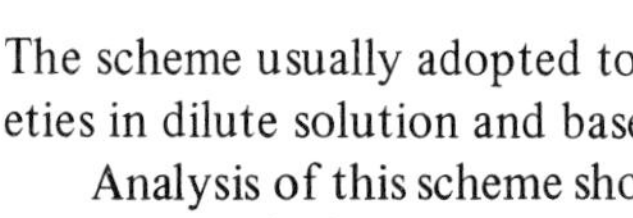

II   Poly(ether sulphone)

I. Simple Excimer Kinetics

The scheme usually adopted to explain excimer kinetics between free aromatic moieties in dilute solution and based upon the notation of Birks[15] is shown in Fig. 18.
    Analysis of this scheme shows that for steady-state excitation, the Stern-Volmer relationship (62) and quantum yield ratio of excimer to monomer fluorescence (63) are obtained.

$$K_{sv} = \left(\frac{\phi_M^0}{\phi_M} - 1 \right) \frac{1}{[M]} = \frac{(k_{FD} + k_{ID}) k_{DM}}{(k_{FM} + k_{IM})(k_{MD} + k_{FD} + k_{ID})} \tag{62}$$

$$\frac{\phi_E}{\phi_M} = \frac{k_{FD}}{k_{FM}} \times \frac{k_{DM}}{(k_{MD} + k_{FD} + k_{ID})} \frac{1}{[M]} \tag{63}$$

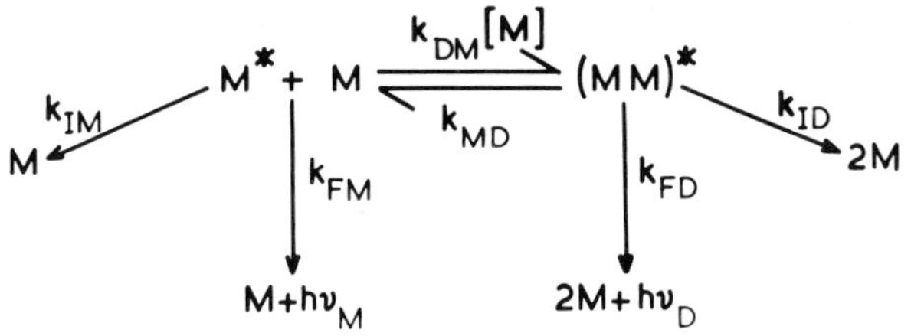

Fig. 18. Birks scheme for excimer kinetics in simple (small molecule) systems

The overall quenching rate constant for monomer fluorescence, k_Q is given by

$$k_Q = \frac{k_{DM}(k_{ID} + k_{FD})}{(k_{MD} + k_{FD} + k_{ID})} \tag{64}$$

Analysis of this scheme for pulsed excitation shows that the fluorescence decay of the uncomplexed aromatic species $i_M(t)$ (referred to here as monomer) is the sum of two exponential terms (Eq. 65) whereas that of the excimer, $i_D(t)$, will be the difference of the same two exponential terms (Eq. 66).

$$i_M(t) = A_1 e^{-\lambda_1 t} + A_2 e^{-\lambda_2 t} \tag{65}$$

$$i_D(t) = A_3(e^{-\lambda_1 t} - e^{-\lambda_2 t}) \tag{66}$$

where

$$\lambda_{1,2} = \frac{1}{2}\,[k_{FM} + k_{IM} + k_{DM}\,[M] + k_{MD} + k_{FD} + k_{ID}$$

$$\pm\,[(k_{FM} + k_{IM} + k_{DM}[M] - k_{MD} - k_{FD} - k_{ID})^2 \tag{67}$$

$$+\,4k_{DM}k_{MD}[M]]^{1/2}]$$

$$A_1 = \left(\frac{k_{FM} + k_{IM} + k_{DM}[M] - \lambda_2}{\lambda_1 - \lambda_2}\right)\,[^1M^*]_{(t=0)} \tag{68}$$

$$A_2 = \left(\frac{\lambda_1 - (k_{FM} + k_{IM}) - k_{DM}[M]}{\lambda_1 - \lambda_2}\right)\,[^1M^*]_{(t=0)} \tag{68}$$

$$A_3 = \frac{k_{DM}[M]}{\lambda_1 - \lambda_2}\,[^1M^*]_{(t=0)}$$

and where

$$\lambda_1 + \lambda_2 = \tau_1^{-1} + \tau_2^{-1} \tag{69}$$

$$= k_{FM} + k_{IM} + k_{DM}[M] + k_{FD} + k_{ID} + k_{MD}$$

We therefore expect monomer decay curves to show dual exponentiality while excimer decays should show a rise followed by an exponential decay.

These kinetics have been established in many small molecule systems, with accurate evaluation of individual rate constants possible when concentration dependence is investigated. Although there have been many apparently successful attempts to fit these simple kinetics to synthetic polymer systems in which fluorescence has been observed, when it is recognised that such polymers are heterogeneous with pendant aromatic groups distributed randomly along the polymer chains, it is perhaps surprising that such simple schemes are successful. Intuitively one might distinguish between two types of monomer species, that which has near neighbouring aromatic groups in such configurations as to enable excimer formation through facile exciton diffusion and small-range conformational changes, and that which would be largely isolated, forming excimers only through long-range interactions. In such cases it is easy to show that additional terms to those given in Eqs. (65) and (66) will be of importance in the fluorescence decay of the polymer systems. One might also distinguish between conformations in the polymers which might lead to excimer fluorescence of differing spectral and decay characteristics. Such mulitple sites would clearly result in kinetics different from those of simple solutions.

A further difficulty is apparent in that whereas the concentration term [M] is well-defined in a free molecule system, it is less easily defined in the case of polymers. This point will be amplified below.

The formation of excimer sites in a vinyl-polymer with pendant aromatic groups must involve rotation in the backbone of the chain, a motion which is clearly hindered[66, 129]. This is illustrated in Fig. 19, from which it can be seen that a single rotation would require a large part of the chain to move through a viscous medium, which would be prohibited on energy grounds. It has been suggested by some that two conformational transitions are correlated in time in a "crankshaft" motion, a process which should require twice the activation energy for the hindered rotation around a single bond in an analogous small molecule, but which is measured to be $12-21$ kJ mole^{-1}, similar to those for butane or 2,4-diphenyl pentane[129]. On the

(a)   (b)

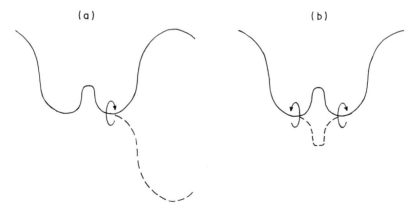

Fig. 19 a, b. Single rotation (a) and concerted "crankshaft" motion (b) in polymer backbones. (after Morawetz[129])

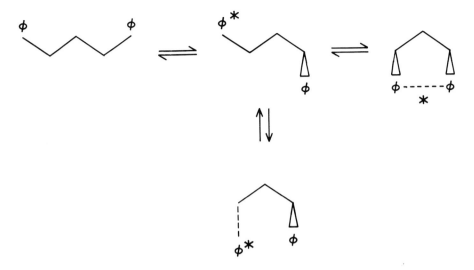

Fig. 20. Excimer formation in 2,4-diphenyl pentane (after Morawetz[129])

basis of observations of excimer formation in 1,3-diphenyl propane summarised in Fig. 20. Morawetz has argued against the importance of "crankshaft" motions in chain molecules such as vinyl polymers, and thus a theory of conformational transitions of such molecules which is compatible with observed low activation energies has yet to be established. Experimental information concerning the motion is readily provided by the study of excimer formation. This is now discussed in various polymers, with emphasis on time-resolved studies.

II. Poly(Styrene)

Excimer fluorescence in solid pure polystyrene is observed exclusively[6, 79, 82, 106, 192]. It has been suggested that in the solid state electronic energy migrates rapidly from the initially excited chromophore to "excimer" trap sites[106], by either exciton diffusion or single-step mechanisms.

In dilute solution, both monomer and excimer fluorescence is observable (Fig. 21)[6, 81, 107, 131, 192, 221].

It has been reported that in ethyl acetate and dichloroethane solution, the position of the excimer band is concentration dependent[131]. The interpretation of solvent effects is complex. Since the compactness of the polymer coil will affect the efficiency of energy migration and the concentration of aromatic species in conformations suitable for excimer formation, solvent effects are to be expected in polymers in which excimer formation is the result of nearest-neighbour interactions[5], as is the case in styrene as shown in studies on styrene-methyl methacrylate copolymers[27].

The relative amounts of monomer and excimer fluorescence are certainly dependent upon the tacticity of the polymer[28, 91].

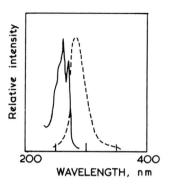

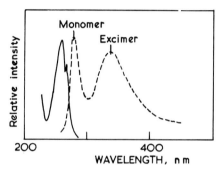

Fig. 21 a, b. Absorption and fluorescence spectra of (a) ethylbenzene, (b) isotactic poly(styrene) in 1,2-dichloroethane solution (after Figs. 2 and 3, Vala et al.[192])

There have been several observations on the time-dependence of fluorescence, summarized in Table 2. It should be noted that the rate constant k_{DM} is given in units of s^{-1}, although in the free molecule scheme the rate constant would be second-order. In polymers M is not a concentration term, but is related to the statistical composition of the polymer (see below).

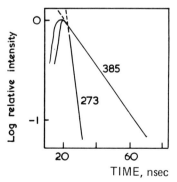

Fig. 22. Fluorescence decay curves for (a) poly(styrene), in CH_2Cl_2 solution. Number denotes emission wavelength monitored; short wavelength corresponds to monomer emission, long wavelength to excimer emission

Table 2. Kinetic data for atactic poly(styrene) fluorescence[a]

Solvent	Φ_M	Φ_D	λ_1^{-1}/ns	λ_2^{-1}/ns	$k_{DM} 10^{-9}$ s^{-1}	$k_{FD} \times 10^{-5}$ s^{-1}	$k_{ID} \times 10^{-7}$ s^{-1}
Cyclohexane[b]	2.7×10^{-3}	9.4×10^{-3}	–	19	2.1	5	5.2
Dioxan	4×10^{-3}	1.5×10^{-2}	<1	22	1.3	6.8	4.5
Methylene chloride	4×10^{-3}	1.6×10^{-2}	<1	15.5	1.3	1.1	6.3
Dichloroethane	4×10^{-3}	1.4×10^{-2}	1.85	17.5	1.3	8.6	5.6
Methylene chloride[c]	–	–	1.02	18.9	–	–	–
Cyclohexane[d]	1.7×10^{-3}	2.6×10^{-2}	–	–	3.4	1.4	5.1

a Based upon Table 7.5[107]
b This polymer is approximately 80% syndiotactic
c Ghiggino et al.[58-61]
d Data for isotactic poly(styrene)

It is of interest to note that within the limit of accuracy of these experiments, monomer decay curves (Fig. 22) were single exponential, whereas Scheme 1 predicts dual exponentiality (Eq. 65). The results thus imply that in polystyrene reverse dissociation ("feedback") of the excimer is not of importance. This point is amplified by time-resolved fluorescence spectra[60] which show that late-gated spectra (see experimental section) are composed exclusively of excimer emission (Fig. 23). The same is true in poly(α-methylstyrene)[60]. In view of more recent work on other vinyl aromatic polymers, it would be of interest to study poly(styrene) further with more sophisticated techniques.

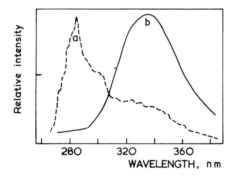

Fig. 23. Gated time-resolved fluorescence spectra of atactic poly(styrene) in dichloromethane solution. (a) Early gated spectrum, delay $\delta t = 0$ ns, gate width $\delta t = 3$ ns. (b) Late gated spectrum, $\delta t = 45$ ns, $\delta t = 3$ ns

III. Poly(Vinyl Naphthalene), Poly(Naphthyl Methacrylate) and Copolymers

As in the case of poly(styrene) in fluid solution both excimer and monomer fluorescence is observed in polymers containing the pendant naphthalene chromophore[2, 5, 30, 107, 131, 155, 156, 173, 192].

In a preliminary study on the time-resolution of fluorescence in poly(1-vinyl naphthalene)[64] the kinetics were constrained to fit Scheme 1, yielding values of monomer decay times in methylene chloride solution of 7.4 and 43.1 ns. Late-gated spectra indicated that reverse dissociation of the excimer occurred. With improvements in techniques, these studies have been greatly amplified recently. In particular, studies on copolymers have permitted more detailed analysis of the "concentration" dependence of excimer formation, and improved statistical analyses have permitted refined modelling of the kinetics. We will discuss at some length one of these papers, and summarize results on other systems.

The series of copolymers of 1-vinyl naphthalene with methyl methacrylate used in a recent study is shown in Table 3[144, 147], where the functions are the mole fraction of naphthalene chromophores in the copolymer, f_n; the fraction of linkages between naphthalene species in the polymer chain, f_{nn}, and the mean sequence length of aromatic species, $\bar{l}_n$. Fluorescence decays in the monomer and excimer emission

Table 3. Copolymer composition data

Copolymer	f_n	f_{nn}	$\bar{I}_n$
1	0.17	0.032	1.15
2	0.27	0.055	1.32
3	0.38	0.124	1.54
4	0.47	0.201	1.81
5	0.58	0.319	2.27
6	0.66	0.421	2.78
7	0.75	0.557	3.95
8	0.83	0.681	5.58

bands were recorded for all eight copolymers at 330 nm and 475 nm respectively. The decay at 330 nm was not adequately described by a dual exponential fit, but satisfactory fits using triple exponential functions of the type:

$$i_{330}(t) = A_1 \exp(-t/\tau_1) + A_2 \exp(-t/\tau_2) + A_3 \exp(-t/\tau_3) \tag{70}$$

were obtained for all copolymer samples.

The deconvolution procedure has been validated for triple exponential fits using synthetic test data. Confidence in the procedure is also enhanced by the high signal to noise ratio obtainable with pulsed laser excitation. It should be noted that dual exponential decay would be anticipated on the basis of Scheme 1.

The excimer decay at 475 nm was not adequately matched by dual exponential fits would be expected according to the conventional monomer/excimer scheme 1. Due to uncertainties with the method in correction for photomultiplier response as a function of wavelength, analysis of excimer kinetics in terms of three exponential terms, although indicated by the data, is not justifiable in any quantitative sense. Decay parameters obtained are given in Table 4.

Table 4. Decay parameters for 1-vinyl naphthalene – methyl methacrylate copolymers

Sample[a]	A_1	τ_1/ns	A_2	τ_2/ns	A_3	τ_3/ns
1	0.005	73.55	0.023	12.65	0.112	40.06
2	0.011	67.60	0.041	15.31	0.092	34.17
3	0.027	59.13	0.057	11.88	0.074	26.44
4	0.021	59.50	0.075	8.46	0.066	23.09
5	0.025	53.34	0.100	6.41	0.069	19.52
6	0.025	53.24	0.115	4.18	0.065	16.36
7	0.032	48.83	0.149	3.57	0.072	14.41
8	0.029	47.07	0.257	1.87	0.072	13.52

[a] Numbers refer to copolymers described in Table 3

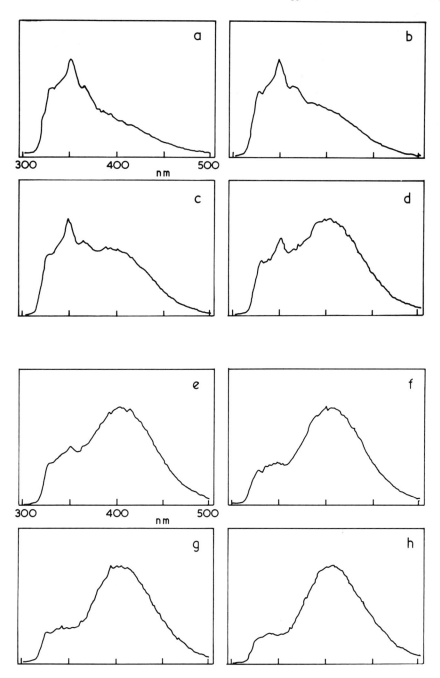

Fig. 24.a–h. TRES for copolymer sample 5 (see text) of vinylnaphthalene-methyl methacrylate series recorded after delays of (a) 0 ns (b) 3.2 ns (c) 6.4 ns (d) 12.8 ns (e) 19.2 ns (f) 25.6 ns (g) 32.0 ns (h) 38.0 ns after excitation

The decay parameters in Table 4 imply the presence of three emitting species. Assuming that conventional monomer and excimer decays are two of the components, it remains to elucidate the nature of the third species. Figure 24 shows gated time resolved emission spectra (Method I in experimental section) recorded for copolymer 5 after a series of time delays, Δt, following excitation. The emission profiles are consistent with the presence of only two spectrally distinct species assigned to monomer and excimer in previous publications concerning steady state and transient excitation studies of polymers containing vinylnaphthalenes. There is little evidence in this work for the presence of a spectrally distinct fluorescent species such as the dimer which has been polstulated[88].

The lifetime data reported in Table 4 were assigned as follows. The long-lived component, τ_1 is expected to be associated with that of the decay of the excimeric species. It might be expected that, if the excimer is populated in part by intramolecular excitation migration from monomer chromophores, the decay time of the latter species would be markedly dependent upon concentration and would diminish at high concentration of aromatic in the copolymer. Consequently τ_2 was assigned to this type of monomer unit. τ_3 was therefore assigned to the third species. Since the chromophore responsible for the decay described by τ_3 was not spectrally distinguishable by time-resolved spectroscopy, and analysis of the data presented in Table 4 indicated that the relative contribution of the emission from the species to the total decreases with increase in aromatic composition, τ_3 was assigned to an unassociated, or monomer, chromophore which can not populate excimer by exciton transfer.

There are two kinetic schemes (2 and 3) which are compatible with these results, shown in Fig. 25.

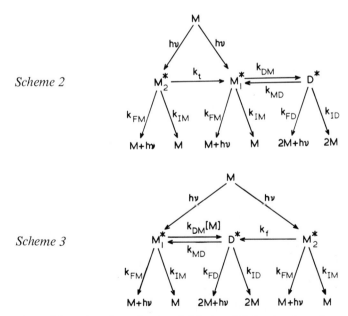

Fig. 25. Possible schemes to explain excimer formation in poly(vinyl-naphthalene) and copolymers. (see text)

Scheme 2. In this photophysical scheme it was proposed that M_1^*, and D^* interact by the generally accepted exciton diffusion mechanism. M_2^* was considered to be an isolated naphthalene chromophore which can transfer energy into M_1^* with a transfer rate characterized by the rate coefficient k_t. Reverse transfer from M_1^* to M_2^* was considered unimportant for the following reason. Exciton diffusion is expected to be very efficient within sequences of naphthalene chromophores within the chain comprising the M_1^* sites. In view of the reduced lifetime of M_1^* relative to M_2^* and of the delocalised nature of the energy within extended chromophore sequences which increases the effective separation of M_1^* and M_2^*, M_1^* to M_2^* energy transfer by Foster or Dexter mechanisms is diminished relative to the M_2^* to M_1^* process.

Scheme 3. An alternative proposal considered M_2^* to be an isolated naphthalene chromophore which can form excimers through a long-range interaction involving segmental diffusion to bring the M_2^* species into close proximity with a ground state chromophore. Such interactions have been cited as the dominant mechanistic routes to excimer formation in polymer such as poly(naphthyl methacrylate). Whilst such interactions have not appeared to dominate the excimer formation in vinylnaphthalene polymers as characterised by steady-state excitation, the possibility of interactions through such a mechanism in addition to nearest neighbour interaction cannot be discounted.

Now the emissions from M_1^* and M_2^* are spectrally indistinguishable and it was assumed that

$$k_M = k_{FM} + k_{IM} \tag{71}$$

was identical for each species, resulting in derivations of decay profiles $i_M(t)$ and $i_D(t)$ recorded for monomer and excimer respectively as

$$i_M(t) = A_1 \exp(-\lambda_1 t) + A_2 \exp(-\lambda_2 t) + A_3 \exp(-\lambda_3 t) \tag{72}$$

and

$$i_D(t) = A_4 \exp(-\lambda_1 t) + A_5 \exp(-\lambda_2 t) + A_6 \exp(-\lambda_3 t) \tag{73}$$

where

$$\lambda_{1,2} = \frac{1}{2} [(X + Y) \pm \{(Y - X)^2 + 4k_{MD} k_{DM} [M]\}^{1/2}] \tag{74}$$

and

$$X = k_M + k_{DM}[M];$$
$$Y = k_D + k_{MD}$$

where

$$k_D = k_{FD} + k_{ID}; \text{ and}$$
$$\lambda_1 + \lambda_2 = k_M + k_{DM}[M] + k_D + k_{MD}; \tag{75}$$

and

$$\lambda_3 = k_t + k_M \text{ (for Scheme (2)) or} \tag{76}$$

$$\lambda_3 = k_f + k_M \text{ (for Scheme (3))} \tag{77}$$

     The empirical triple exponential fit to $i_{330}(t)$ is thus consistent with the expression for $i_M(t)$ in Eq. (72) derived from either of the kinetic schemes 2 and 3 outlined above. Furthermore, although fitting of a triple exponential function to $i_{475}(t)$ was not considered meaningful the inadequacy of dual exponential functions in description of $i_{475}(t)$ offers support for the existence of an excimer decay, $i_D(t)$, the temporal characteristics of which may be described by Eq. (73).

Determination of Rate Coefficients
For low molar mass species in which intermolecular excimer formation results from a diffusion controlled interaction, individual rate parameters may be determined by the following methods.

1. Concentration dependence studies. From a study of the concentration dependence of λ_1 and λ_2 rate parameters may be extracted from the empirical data by a variety of extrapolation techniques.
a) k_M may be estimated from the unquenched monomer lifetime.
b) Since $\lambda_1 \rightarrow k_M$ as $[M] \rightarrow 0$, k_M may be estimated from the intercept of λ_1 as a function of $[M]$.
c) Since $\partial \dfrac{(\lambda_1 + \lambda_2)}{\partial [M]} = k_{DM}$, k_{DM} is conveniently estimated as the slope of a plot of $(\lambda_1 + \lambda_2)$ against $[M]$.
d) As $[M] \rightarrow \infty$, $\dfrac{\partial \lambda_2}{\partial [M]} \rightarrow k_{DM}$.
k_{DM} may be obtained, as an alternative to method (c), from a plot of λ_2 as a function of $[M]^{-1}$.
e) Since $(\lambda_1 + \lambda_2) \rightarrow k_M + k_{MD} + k_D$ as $[M] \rightarrow 0$ the intercept of $(\lambda_1 + \lambda_2)$ against $[M]$ may be used in combination with (a) or (b) to estimate $(k_{MD} + k_D)$.
f) Since $(\lambda_1 \lambda_2) \rightarrow k_M(k_{MD} + k_D)$ as $[M] \rightarrow 0$, $(k_{MD} + k_D)$ may be estimated as an alternative to method (e) from a plot of $(\lambda_1 \lambda_2)$ as a function of $[M]$ through substitution of k_M from (a) or (b).
g) The slope of $(\lambda_1 \lambda_2)$ vs M furnishes $k_{DM} k_D$. Thus k_D may be estimated using the value of k_{DM} from either (c) or (d).
h) As $[M] \rightarrow \infty$, $\lambda_1 \rightarrow k_D$.
i) k_{MD} may be estimated by combinations of (e) and (f) with (g) and (h).

2. Single concentration approach. An alternative, less rigorous, technique is to employ kinetic data derived for a single chromophore concentration in combination with the unquenched lifetime[15]. Any attempt to derive equivalent rate coefficients in homopolymers must of necessity be confined to the less rigorous single concentration approach (2)[5, 94]. This inevitably produces a pseudo first order "k_{DM}" term which is of the form $k_{DM}[M]$ where $[M]$ is, for a given polymer, a constant but undefined local concentration of chromophore within the polymer coil. A modified version of the single concentration technique involving transient decay measurements in the presence of dynamic quenching agents has produced rate parameters which suffer the same limitation[92, 173].

The use of copolymers incorporating aromatic species offers a unique opportunity to determine the rate parameters governing intramolecular excimer formation in macromolecules using the extrapolation techniques outlined in method (1) above. Furthermore it is possible, in principle, to remove the concentration dependence from the rate coefficient assigned to excimer formation. The problem encountered in the use of copolymers is the selection of an appropriate function to describe the concentration of chromophores which occupy potential excimer sites in the molecule. The formulation of appropriate concentration terms have been discussed with respect to description of relative excimer to monomer emission efficiencies in steady state conditions[2, 3, 154–156]. The concentration term must account not only for the geometric considerations implied in the excimer formation process but also for the extent to which migration is capable of populating such sites.

Analysis of steady state fluorescence of polymers containing 1-vinylnaphthalene resulted in the choice of the term $f_{nn} \tilde{l}_n$ to describe the intramolecular concentration dependence of the observed excimer characteristics. The same concentration function was thus adopted in the time-resolved work and produced more satisfactory fits to concentration dependences of functions concerning λ_1 and λ_2 (as plotted in Figs. 26 to 29) than did any other terms descriptive of the chain microcomposition.

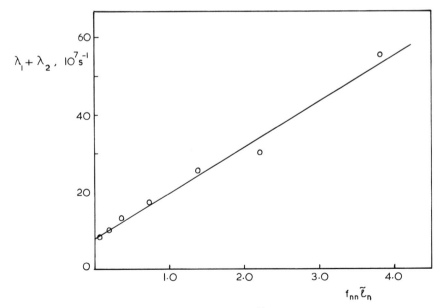

Fig. 26. Plot of $(\lambda_1 + \lambda_2)$ (see text) as function of $f_{nn} \tilde{l}_n$ for methylmethacrylate -1 vinyl naphthalene copolymers

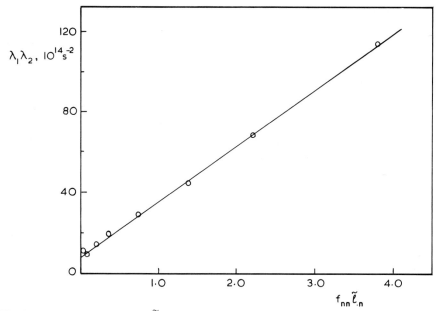

Fig. 27. Plot of $\lambda_1 \lambda_2$ against $f_{nn} \tilde{l}_n$ for same polymers as in 26

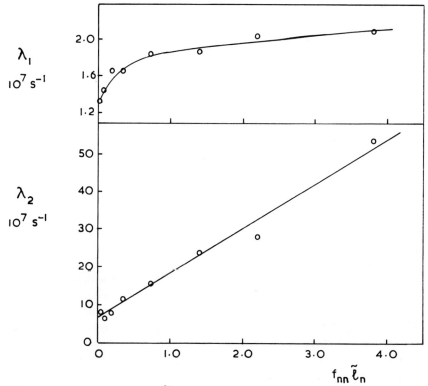

Fig. 28. Plot of λ_1 and λ_2 against $f_{nn} \tilde{l}_n$ for polymers as in 26

Analysis of schemes 2 and 3 reveals that the existence of M_2^* does not affect the significance of λ_1 and λ_2 relative to the scheme 1 provided that reverse activation of M_2^* via k_{-t} or k_{-f} is insignificant.

Consequently, the modification of method (1) for the determination of the rate parameters descriptive of intramolecular excimer formation in copolymers containing vinylnaphthalene is justified. It is important to note however that since the pre-exponential terms A_1 and A_2 are altered by the presence of M_2^*, the presence of such a species will invalidate application of the single concentration technique (method (2)) to polymers in which dual exponentiality in the decay analysis is not observed.

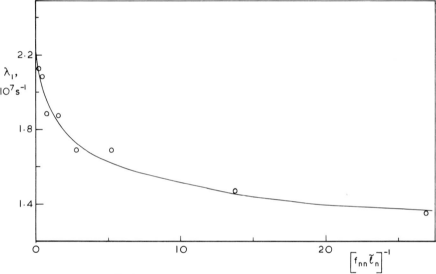

Fig. 29. Plot of λ_1 vs $f_{nn} \tilde{l}_n^{-1}$ for these polymers (see text)

The functions incorporating λ_1 and/or λ_2 necessary for generation of decay parameters as outlined in method (1) above are plotted as a function of the concentration term $f_{nn} \cdot \tilde{l}_n$ in Figs. 26 to 29. The linearity of the Figs. 26 and 27 over the entire copolymer concentration range studied lends credibility to the validity of adoption of the concentration term $f_{nn} \cdot \tilde{l}_n$ and to the applicability of the analytical procedure. The resultant kinetic parameters are given in Table 5.

It is significant that the value of k_M of 1.3 (± 0.1) x 10^7 s^{-1} obtained by the extrapolative technique of method 1 (b) is in poor agreement with the values of 1.72 (± 0.01) x 10^7 s^{-1} and 2.15 (± 0.05) x 10^7 s^{-1} determined from the single exponential decays observed using 2 x 10^{-5} M 1-methylnaphthalene and a copolymer containing less than 0.5 mole % Naphthalene chromophore respectively. The value of 1.7 x 10^7 s^{-1} obtained for the 1-methylnaphthalene is in excellent agreement with that of 1.8 x 10^7 s^{-1} for 1-ethylnaphthalene in 2-methyltetrahydrofuran at 25 °C[95]. This implies that care must be exercised in choice of model compounds if the single concentration method (2) is to be adopted in these studies.

Table 5. Kinetic parameters

Kinetic parameter	Value/10^7 s^{-1}	Method
k_M	1.72	1.a)[a]
	2.15	1.a)[b]
	1.3	1.b)
αk_{DM}	12.1	1.c)
	12.0	1.d)
$k_{MD} + k_D$	6.5	1.e)
	6.2	1.f)
k_D	2.2	1.g)
	2.2	1.h)
k_{MD}	4.2	1.i)

[a] Determined using 2×10^{-5} m 1-methylnaphthalene
[b] Determined using copolymer containing less than 0.5 mole % naphthalene chromophore

It should further be noted that although copolymer studies offer an advantage over single concentration studies an absolute value of k_{DM} still can not be obtained. Since the concentration term $f_{nn} \cdot \tilde{l}_n$ is proportional to the "true" molar concentration of chromophore within the polymer coil i.e.

$$M = \alpha f_{nn} \tilde{l}_n \tag{78}$$

the excimer formation rate coefficient as determined in this method corresponds to αk_{DM}. However, the value αk_{DM} is an appropriate rate parameter to describe excimer formation since it is not dependent upon variations in intramolecular chromophore concentration, α merely being a numerical constant.

These studies have shown that there are essentially two kinds of monomer species in these vinyl naphthalene polymers with differing capacity for excimer formation, whereas in earlier studies[58-60, 94, 95] only one monomer species was observed. However, examination of Table 4 reveals that τ_2 becomes very short at high aromatic contents in the copolymers. It is therefore likely that the component of this decay will be extremely fast in the homopolymer, as expected for a mechanism involving excimer site population by exciton diffusion. Consequently the lifetimes measured on a nanosecond timescale from poly(vinyl naphthalene) analysed in terms of dual exponential fits will be weighted averages dominated by τ_1 and τ_3, if τ_2 is below the instrument resolution. This would explain the occurrence of two long lived decays in polymers in which a sub-nanosecond monomer decay time would have been expected on the basis of the conventional kinetic scheme. Moreover, the measurement of two decay parameters, *approximating* to λ_1 and λ_3 rather than to λ_1 and λ_2 for scheme 1 would alter the interpretation of the rate coefficients derived using the single concentration approach method (2). Thus the value of the concentration dependent "k_{DM}" derived in such a technique would relate to the formation of excimer form M_2^* characterized by k_i in the work described above.

Some further evidence on the nature of the third emitting species M_2^* has been obtained from studies on 1-vinyl naphthalene-methyl acrylate copolymers[145] and an as yet unreported temperature study. Rate-constants obtained from the more flexible acrylate copolymers obtained by an analysis analogous to that described above are given in Table 6. It is of interest that rate parameters for the different co-polymers which are internal deactivation rates of monomer (k_M) and excimer (k_D) agree well. Moreover rate parameters which might be expected to depend significantly on the flexibility of the polymer chains show the anticipated dependence, in that αk_{DM}, and k_{DM} are much greater in the more flexible acrylate copolymers, than in the methacrylate polymers.

For both sets of copolymers, and in either scheme 2 or scheme 3, λ_3 is of the form;

$$\lambda_3 = k_M + k_i \tag{79}$$

where $k_i = k_t$ or k_f depending upon the reaction scheme. Furthermore, λ_3 was found to be linearly dependent upon the mole fraction of naphthalene derivative in the co-polymers, f_n, as demonstrated in Fig. 30. Since k_M is independent of concentration k_i must be of the form.

$$k_i = k_i' f_n \tag{80}$$

Such a concentration dependence would be consistent either with a Foster or Dexter transfer mechanism whose rate is such that Stern-Volmer kinetic behaviour is obeyed or would be consistent with the alternative suggestion of scheme (3) where the excited state naphthalene chromophore experiences intramolecular diffusion con-trolled self quenching.

Table 6. Kinetic data for 1-vinyl naphthalene – methyl acrylate copolymers[145]

Rate Coefficient	Value x $10^{-7}/s^{-1}$	Procedure
k_M	1.91	a)[a]
	1.72	a)[b]
	1.5	b)
αk_{DM}	46.5	c)
	43.0	d)
$k_{MD} + k_D$	14.5	e)
	14.0	f)
k_D	2.36	g)
	2.35	h)
k_{MD}	11.95	i)

[a] Determined using 2×10^{-5} M 1-methylnaphthalene
[b] Determined using copolymer containing less than 0.5 mole % chromophore

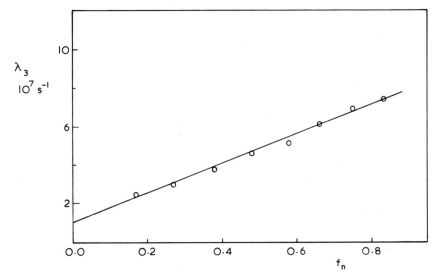

Fig. 30. λ_3 vs f_n for the same polymer series

Comparison of the value of k_i' of 11.4 $(\pm 0.5) \times 10^7$ s^{-1} for the 1-vinylnaphthalene/methyl acrylate copolymer series with that of 7.8 $(\pm 0.3) \times 10^7$ s^{-1} obtained for the 1-vinylnaphthalene/methyl methacrylate system show a slight dependence of the rate coefficient upon chain flexibility. Since dielectric relaxation measurements upon poly(methyl methacrylate) and poly(methyl acrylate) in solution indicate that the segmental motion of the latter system is an order of magnitude faster at 298 K[230] it is tempting to assume that the mechanism outlined in Scheme (2) is more appropriate than that of Scheme (3) which requires long range interactions via polymer chain coiling. However, the situation is complicated by the considerable modification of polymer flexibility induced by vinyl-naphthalene chromophores, particularly at high concentrations. Furthermore, fluorescence depolarization measurements of the relaxation of PMMA and PMA bearing fluorescent probes of differing molar volume indicate[135] that the difference in relaxation rate may not be as marked as indicated by dielectric relaxation techniques. A temperature study shows that the activation energy associated with k_i is small, of the order of 2 kJ mol^{-1} [146] and this seems incompatible with chain coiling. Thus Scheme 2 may be the preferred explanation.

It will be noted that the kinetic scheme adapted above reverse dissociation of excimer sites, the experimental evidence for this being the triple exponential decay of the monomer fluorescence and the gated time-resolved fluorescence spectra, reported here and seen earlier[58-61, 87] in the homopolymer.

A recent study[167] on poly(2-vinyl naphthalene) P 2 VN has also concluded that the simple scheme I is inappropriate for the description of excimer kinetics in this polymer, and on the basis of comparison with isomers of 1,1'-di-2-naphthyldiethyl ether (III), explanations of behaviour which emphasize conformational aspects (see[16, 54, 76]) have been proposed. In the polymer, monomer fluorescence decays monitored at 330 nm were observed to be multiexponential, an observation similar

CH$_3$—CH—O—CH—CH$_3$

III

to that in poly 1-vinyl naphthalene discussed above. However, the excimer fluorescence in P2VN was reported as being triple-exponential. This was qualitatively described in P1VN, but not analysable with sufficient accuracy to be reported in that polymer[144]. Late gated spectra in P2VN apparently showed only excimer emission, in contrast to those from P1VN[60, 87, 144] in which a monomer component is clearly visible.

The meso and racemic isomers of III Fig. 31 exhibit some important differences in fluorescence behaviour. In the racemic compound excimer intensity is smaller than that in the meso, and the I_D/I_M ratio in the meso compound is wavelength dependent, whereas this is not the case in the racemic compound. The temperature dependencies of emission are also different. At temperature below 190 K, the meso compound analysed at 430 nm (excimer region) exhibits a growing in and dual-exponential decay, whereas above this temperature the sum of two exponentials suffices (cf. P1VN discussed above). It was thus concluded that the dual exponential decay of the excimer in P2VN could be explained on the basis of contributions of isotactic sequences to total excimer intensity. Thus the meso form of III with an almost exclusive population in the TG conformation could form the TT excimer at a fast rate with a low energy barrier. The dual exponentiality in excimer emission observed could be explained if naphthyl groups in the ground state an adopt different orientations prior to excimer formation. The formation of the GG excimer could explain the dual exponential nature of the decay, but is less likely on energetic grounds.

One of the possible explanations for the kinetics of racemic III fluorescence (in which dual exponential decay in the monomer region is observed) is shown in Scheme 4 Fig. 32, in which the establishment of an equilibrium between GG and TT ground-state and excited state conformation precedes formation of a TG excimer from TT*. Intramolecular excimer formation in dinaphthyl alkanes has been similarly shown to be related to the conformation of the ground-state molecule and the facility for conformational relaxation to the excimer state[133]. Thus k_{DM} (see Scheme 1) for meso 2,4-dinaphthyl pentane was much greater than that for racemic 2,4-dinaphthyl pentane, as in III above.

It is clear that the time-resolved studies of those polymers are in an interesting and productive phase. With further work it should be possible to rationalize the kinetically distinguishable monomer and excimer species in terms of the different conformation sites within polymers.

There have been similar time-resolved studies recently upon naphthyl methacrylates, IV and V[84].

Once again, analysis of the fluorescence kinetics of IV and V in terms of Scheme 1 is not possible. For the monomer, dual exponential decay was sufficient to describe the time dependence of the fluorescence, the longer component not matching the long decay in the excimer (Table 7[84]), but resembling closely the decays of model

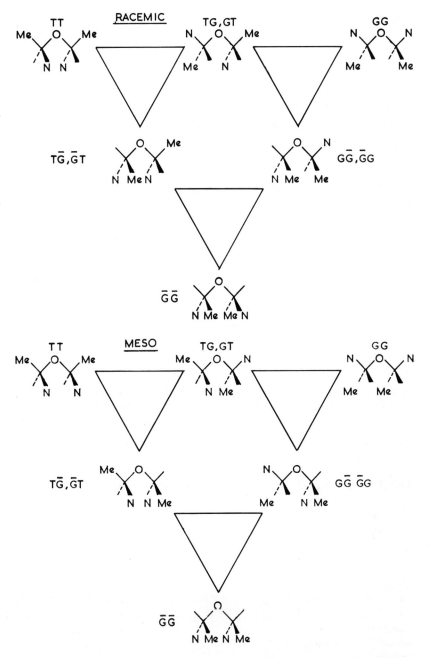

Fig. 31. Structures of 1,1′-di-2-naphthyl diethyl ether. (After de Schryver et al.[167])

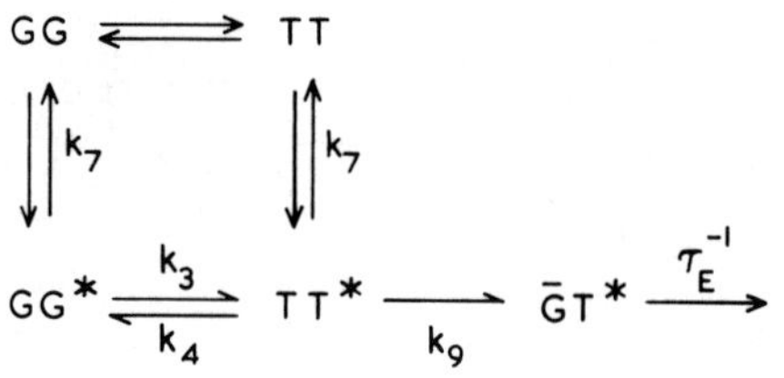

$$GG \rightleftarrows TT$$

$$\Big\uparrow\Big\downarrow k_7 \qquad \Big\uparrow\Big\downarrow k_7 \qquad\qquad \textit{Scheme 4}$$

$$GG^* \underset{k_4}{\overset{k_3}{\rightleftarrows}} TT^* \xrightarrow{k_9} \bar{G}T^* \xrightarrow{\tau_E^{-1}}$$

Fig. 32. *Possible* explanation of excimer formation in 1,1'-di-2-naphthyl diethyl ether. (After de Schryver et al. [167])

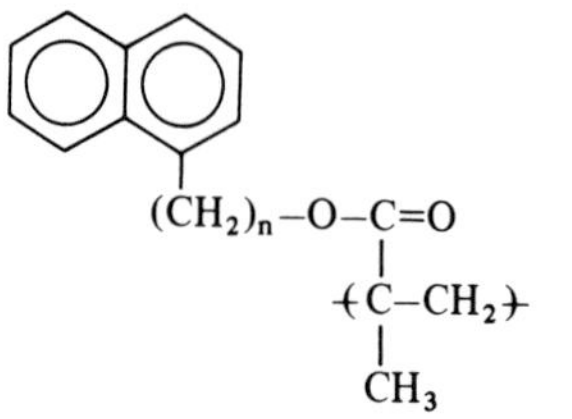

$$(CH_2)_n-O-C=O$$
$$\underset{\underset{\displaystyle CH_3}{|}}{\overset{|}{+C-CH_2+}}$$

IV n=1
V n=2

compounds. Monomer fluorescence can thus be separated into two contributions, one arising from an isolated chromophore which behaves like the model compound, the other with excimer-forming properties. These observations are very similar to those on P1VN discussed at length above with the difference that in IV and V no reverse dissociation of the excimer is observable. A rather elegant demonstration of the existence of an isolated monomeric unit was given by dispersing IV, V and P1VN in poly(styrene) or poly(methyl methacrylate) matrices[84]. The same long lived monomer component is observed as in solution, again recognisably different from the long-

Table 7. Components of fluorescence decays of monomer and excimers of IV and V[84]

Sample	Monomer τ_2/ns	Excimer τ_2/ns
IV[a]	36	48
IV[b]	40	63
V[a]	66	60
V[b]	58	76
Model[a, c]	34	
Model[b, c]	39	
Model[a, d]	63	
Model[b, d]	60	

[a]　tetrahydrofuran solvent
[b]　toluene
[c]　Model is 1-naphthyl methyl pivalate
[d]　Model is 2(1-naphthyl) ethyl pivalate

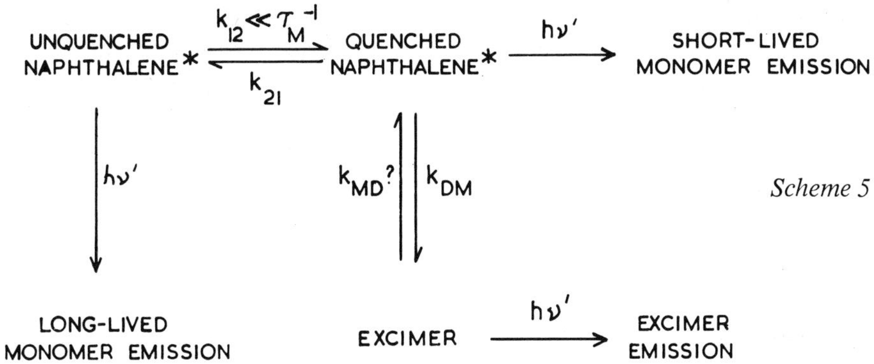

Scheme 5

Fig. 33. Scheme proposed to explain excimer formation and decay in naphthyl methacrylates. (After Holden et al.[84])

lived excimer decay. The results on IV and V were rationalised on the basis of scheme 5, (Fig. 33) which resembles schemes 2 and 3 closely, although the interpretation of Holden et al. that the isolated chromophores are "unquenched" is not borne out by studies on copolymers.

IV. Poly(Acenaphthylene) and Copolymers

The principal difference between considerations of fluorescence behaviour of acenaphthylene polymers VI and those of the photophysical characteristics of vinylaromatic polymers is that whilst the latter systems may form excimers through interactions between nearest-neighbour chromophores on the polymer chain, steric restraints preclude such a mechanism in polymers of acenaphthylene[26, 210]. It has been suggested that the dominant mechanism for excimer formation in acenaphthylene derived systems involves interactions between next to nearest neighbours. Such a proposal has been validated for copolymers of acenaphthylene with methyl methacrylate[156] and methyl acrylate[3], and is reinforced by the observation of excimer emission in alternating copolymers incorporating acenaphthylene chromophores[210] in steady state excitation.

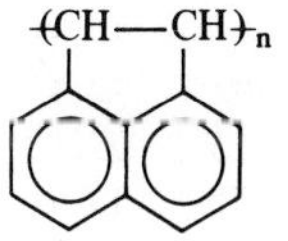

VI

   Once again, time-resolved studies on copolymers have shown the existence of two kinetically distinct monomer species, an "isolated" chromophore capable only of excimer formation through long-range interactions; and a monomer capable of excimer formation which is also populated by reverse dissociation[144, 147]. An alternative explanation of observed behaviour is that two excimer species may exist which are populated from a single excited monomer chromophore and which display different decay rates. Such a mechanism has been proposed in description of the kinetic behaviour of polymers containing N-vinylcarbazole (see below). However time resolved fluorescence spectra of acenaphthylene polymers indicate the presence of only two spectrally distinct species. Therefore the existence of two energetically dissimilar excimer states is precluded. The only other possible situation corresponding to case (i) above would involve excimer states of similar spectroscopic nature but of differing temporal character. Such a situation might be envisaged in poly(acenaphthylene) since different steric constraints within excimers of the type AMA and AAA (where M = methyl methacrylate and A = acenaphthylene might lead to different kinetics of association and dissociation. This possibility was discounted on the following grounds.

a) Time-resolved spectroscopy revealed that the excimer state was longer lived than that of the monomer. Consequently the short decay time τ_2 may be assigned to that of a monomeric emission. Since late gated emission spectra are dominated by excimer emission, τ_1 must correspond to that of an excimer. At high methyl methacrylate concentrations the excimer sites are almost entirely AMA. At low concentrations the excimer emission will contain an increasing contribution from AAA sites. Since there are no obvious spectral changes there is no evidence for the existence of two time-resolved excimers within the decay characterized by τ_1.

b) If τ_1 *and* τ_3 originated from the influence of two time-resolved excimers there is no reason to suppose that τ_3 should vary with chromophore concentration in the manner evident from the data in Table 8.

Schemes 2 or 3 could again be used to model the kinetic behaviour in this polymer, yielding rate-constants shown in Table 9.

Examination of the data in Table 8 reveals that τ_3 (and hence λ_3) is a function of the intramolecular chromophore concentration. In previous studies of vinylnaphthalene copolymers λ_3 has been found to vary linearly with the mole fraction of aromatic, f_a. Figure 34 demonstrates the reasonable fit attained in such a plot for the acenaphthylene/methyl methacrylate system. However, unlike the vinylnaphthylene copolymer systems, the linear fit of λ_3 to f_a as a concentration term is not unique.

Table 8. Decay parameters for $i_{330}(t)$ for copolymers of acenaphthylene with methyl methacrylate[147]

Sample	$f_a{}^a$	Sample[b] $f_a \Sigma$	A_1	τ_1/ns	A_2	τ_2/ns	A_3	τ_3/ns
2	0.090	0.016	0.023	34.1	0.011	7.80	0.094	34.3
3	0.200	0.086	0.021	35.1	0.033	6.81	0.126	22.1
4	0.300	0.169	0.023	31.7	0.040	5.19	0.085	17.3
5	0.400	0.268	0.019	35.3	0.087	4.31	0.126	14.3
6	0.460	0.343	0.016	35.9	0.101	3.97	0.095	13.7
7	0.550	0.410	0.024	33.8	0.138	3.42	0.110	11.3
8	0.630	0.434	0.025	34.7	0.169	3.57	0.110	11.4
9	0.730	0.457	0.022	33.9	0.174	3.33	0.092	11.4

[a] f_a is mean mole fraction of aromatic in the copolymer
[b] Σ is the pentad distribution function[156]

Table 9. Rate-constants for decay in acenaphthylene – methyl methacrylate copolymers.[147]

Rate coefficient[a]	Value/10^7 s^{-1}
k_M	2.88
	3.0
αk_{DM}	39.0
	39.1
$k_{MD} + k_D$	12.2
	12.2
k_D	2.9
	3.0
k_{MD}	9.3
k_i'	11.3[b] (12.9)[c]

[a]   As defined in Schemes 2 and 3
[b]   From Fig. 33
[c]   From Fig. 34

An equally acceptable fit is found for λ_3 as a function of $f_a \cdot \Sigma$ as demonstrated in Fig. 35. Whichever fit is appropriate, λ_3 may be described by the following relationship

$$\lambda_3 = k_M + k_i \tag{81}$$

where $k_i = k_t$ or k_f according to the relevant reaction scheme 2 or 3. The concentration dependence of k_i may be expressed as

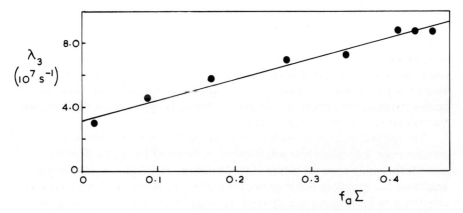

Fig. 34. Plot of λ_3 vs $f_a \Sigma$ (see text) for poly acenaphthylene copolymer series

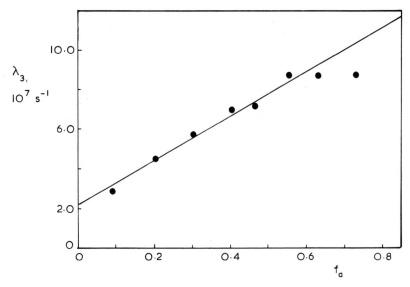

Fig. 35. Plot of λ_3 vs f_a (see text) for same polymer series as in Fig. 34

$$k_i = k_i'[M] \tag{82}$$

where [M] represents some concentration term representative of the intramolecular concentration governing the mechanism responsible for the existence of the second monomeric component. Consequently k_M may be obtained by extrapolation of a plot of λ_3 against [M] to [M] = 0. The resultant values of k_M derived from the plots shown in Figs. 34 and 35 are also listed in Table 9. The use of either of the concentration terms f_a or $f_a\Sigma$ produces values of k_M which are in reasonable agreement, given the error in extrapolation, with those detailed in Table 9.

As discussed above a dependence of λ_3 upon f_a is consistent with either photophysical scheme 2 or 3. However, if the energy migration in acenaphthylene polymers is the result of a series of Foster interactions between chromophores distributed within the polymer coil (as deduced from studies of the depolarization of emission as a function of f_a [156]) it is not immediately apparent to the authors how a reservoir of isolated monomer units M_2^* can be established which can not interact in the conventional excimer formation mechanism. Further information on the nature of these interactions and upon the overall validity of kinetic schemes could be furnished from studies involving copolymers of differing coil flexibility and microcomposition, and through variation of temperature and solvent compatibility.

The fact that excimer formation cannot occur as a result of nearest neighbour interactions in acenaphthylene polymers and of the possibility of the dependence of excimer formation upon the nature of the "bridging" molecule in next to nearest neighbour configurations makes analysis of the kinetic data in these polymers more complex than for vinylaromatic systems. This complexity implies that the approximations involved in the kinetic treatment of David et al.[31] in which the influence of excimer dissociation is neglected and the decay time of excimer emission equated to k_D^{-1} for all intramolecular concentrations may be invalid.

V. Poly(N-Vinyl Carbazole) and Copolymers

In recent years the fluorescence from poly(N-vinylcarbazole) has received much attention, partly due to its photoconducting properties and partly due to the unique photophysical processes which it displays. In contrast to other aromatic containing polymers, no emission from monomer — like moieties has been observed[100]. Even in solution the broad unstructured fluorescence profile is resolved into two spectrally distinct excimer sites[61, 87, 93, 100]. One of these, with emission maximum at 420 nm, has, from studies on model compounds, been attributed to the normal "sandwich-type" excimer formed between neighbouring carbazole groups on the polymer chain in a totally eclipsed conformation[100]. The structure of the other excimer, with emission maximum at 380 nm, has been less well characterized although Johnson has suggested a dimeric structure with considerable deviation from coplanarity, of the two carbazole rings. Itaya et al.[93] on the other hand, have proposed a structure with only one eclipsed aromatic ring from each group. Interconversion between this species and the low-energy "sandwich-type" excimer has been demonstrated using time-resolved fluorescence spectroscopy[58-61, 87].

The temperature dependence of the fluorescence from a solution of poly(N-vinylcarbazole) has been shown to be unusual[100, 159] since the isoemissive point expected for an interconverting two-component system is absent. Instead the fluorescence due to the low energy excimer remains essentially invariant, whilst that for the high energy species displays a marked temperature dependence having maximum emission at low temperatures (Fig. 36). In that report, also, it was again not possible to describe the fluorescence decay using a two component decay law, as may have been expected for two interconverting species.

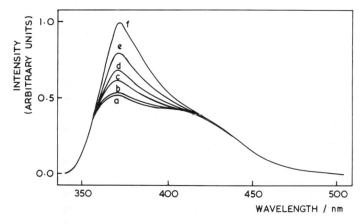

Fig. 36. Temperature dependence of emission intensities from excimer sites in poly(N-vinyl carbazole). a = 295, b = 275, c = 260, d = 240, e = 220, f = 200 K

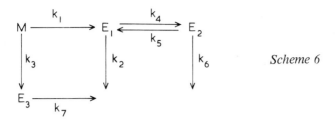

Fig. 37. Kinetic scheme for formation and interconversion of excimer sites in poly(N-vinyl carbazole)

More recently, the fluorescence from a solution of poly(N-vinylcarbazole) has been resolved into three components[157, 158, 184] and the kinetic scheme 6 has been proposed (Fig. 37).

The two reports differ only in the interpretation of E_1. Tagawa et al.[184] suggest a "relaxed monomeric" group whilst Roberts et al.[157-159], on the basis of spectral evidence, prefer a structure similar to E_3 but in such a configuration in the polymer chain as to be able to form E_2 reversibly. Three decay times of 2.5 ns, 34.9 and 11.5 ns were assigned to E_1, E_2 and E_3 respectively at room temperature in dilute solution[157-159]; τ_2 is largely independent of temperature whereas τ_1 and τ_3 are strongly temperature dependent[158]. The "second excimer" species, characterized by emission at 380 nm (E_1 and E_3) is formed in less than 10 ps[184], and must therefore arise from preformed excimer sites, as Johnson suggested. The sandwich excimer at 420 nm is formed, however, on a nanosecond time-scale. The presursor to the sandwich excimer is either a relaxed monomeric species[184] or a fraction of the "second" excimer sites. There may of course be a whole distribution of second dimer sites of differing geometry and thus decay parameters (although not apparently spectral characteristics), but within the experimental limits set by curve-fitting procedures, these resolve themselves into only two *kinetically* distinguishable components, E_1 and E_3.

The relative contributions to the fluorescecne of these preformed sites E_1 and E_3 as a function of temperature has been evaluated in a recent temperature study[59] (Fig. 38), which shows that Φ_1 is reduced whilst Φ_3 is increased dramatically as the temperature is lowered. The large change in Φ_3 is not consistent with the small observed variation in τ_3 unless the branching ratio $k_3/(k_1 + k_3)$ is temperature dependent, possibly due to a change in the ratio of performed E_1/E_3 sites, E_3 being favoured at low temperatures. This is consistent with the observation of a 15 ns decay time reported by Johnson[100] for the same system at 77 K.

The behaviour of poly(N-vinyl carbazole) is certainly complex, and an attempt has been made recently to gain an insight into the photophysical properties of this polymer by studying the related compounds below[102]. Monomer fluorescence is observed in these compounds

VII   Poly (N-ethyl-2-vinyl carbazole)
VIII   Poly (N-ethyl-3-vinyl carbazole)
IX   Poly (N-ethyl-4-vinyl carbazole)
X   1,3-bis (N-carbazolyl)propane
XI   1,3-bis (N-ethyl-2-carbazolyl)propane

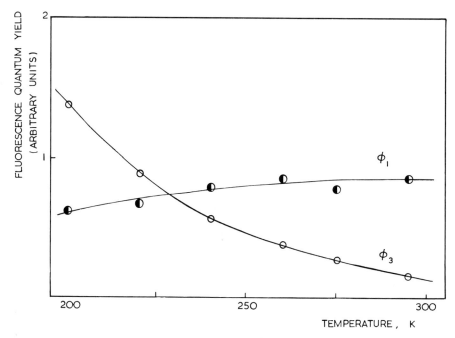

Fig. 38. Temperature dependence of quantum yields of fluorescence from sites E_1 and E_3 (see Fig. 37 and text) in poly(N-vinyl carbazole)

Table 10. Decay of carbazole-containing polymers in THF, 25 °C[102])

Polymer	τ_M	τ_E	$k_{DM} \ s^{-1}$
VII	8	22	6.4×10^7
VIII	3.2	17	2.4×10^8
IX	4	11	1.7×10^8
X	6.4	15.9	9.6×10^7
XI	10.8	14.7	3.2×10^7

For all compounds, decays were fitted to Scheme 1, although procedures followed for doing this were unsophisticated and results may be uncertain. The approximate values obtained for monomer (τ_M) and excimer (τ_E) are given in Table 10. This study shows quite clearly the importance of kinetic measurements. Thus whereas the fluorescence spectra of VII and IX are indistinguishable, the rate parameters differ considerably. These differences undoubtedly reflect the variations in preferred ground state geometry between neighbouring pendant groups determined by the position of bonding to the polymer backbone. In IX, only a minor conformational change is required to form a sandwich excimer, whereas for example VII requires a much more substantial change.

VII   VIII

IX

It is hoped that this section has illustrated the applicability of time-resolved fluorescence methods to the study of conformational motion of synthetic polymers and given an indication of the recent but increasing interest in this field.

D. Time-Resolved Studies on Biopolymers

This is a field of increasing activity, and it would be quite impossible to give a comprehensive review within the confines of this article of all papers dealing with fluorescence from molecules of biological interest. Instead we discuss here those recent papers judged to be of importance in which time-resolution has played a critical role in furthering the understanding of the molecular systems. It should be noted that excimers have been used as probes in biological membranes also[166] and (Sect. E below).

I. Information from Steady State Measurements

The steady state luminescence from naturally occurring macromolecules has been extensively studied in the past to investigate the effect of light on the properties of biological systems and to reveal some aspects of conformation and structure (see reviews[111, 120, 179]). Certainly much useful information has been obtained about the type and behaviour of excited state of biopolymers from such measurements. However, the complex and heterogeneous nature of these systems have often led to ambiguities in interpretation of the results. The fluorescence from proteins, for example, has largely been attributed to the constituent aromatic amino acids phenylalanine, tyrosine and tryptophan. Many workers have used the observation that the emission characteristics of these amino acids are sensitive to the environment surrounding the residue (e.g. polarity, microviscosity, presence of quenching groups) to draw conclusions about the protein structure near the emitting site[228, 231]. Thus observed variations in fluorescence maxima, quantum yield and lifetime of tryptophan residues in proteins have led to the assignment of three discrete types of residue environment[231]. In other work the emission properties of various fluorescent probes bound to proteins, lipid bilayers and biological membranes have been used as indicators of environment character and function[25, 190, 224, 232]. Steady state luminescence measurements, however, provide only a limited amount of information as they represent the *total* contribution of emission processes occurring during the excited state lifetime. It has been conclusively shown by Weber and others[38, 71, 215] that fluctuations in the structure of biopolymers occur on a time scale comparable to the fluorescence process and lead to complex spectral and temporal luminescence behaviour. The application of time-resolved fluorescence techniques promises to provide a greater insight into the interesting dynamic behaviour of biological systems and some recent results and the future possibilities of such measurements are considered below.

II. Dynamic Excited State Properties of Proteins and Nucleic Acids

The luminescence properties of polynucleic acids[41] and proteins[228] have been reviewed in detail previously. However recent time-resolved studies have important implications for existing concepts of the excited states of biopolymers. The readily observable luminescence from proteins at room temperature has made these systems particularly attractive to study. The fluorescence properties of the aromatic amino acid zwitterions which determine the emission from proteins are summarised in Table 11. The wide range of emission maxima observed in proteins (ranging from 308 nm in azurin to 342 nm in lysozyme) has been explained by many authors to reflect the different environments of the tryptophan residues in the proteins[231, 233]. However, it is now clear that such measurements do not give an unambiguous indication of structural differences in the residue environment. Several workers have measured the fluorescence decay kinetics of a number of proteins and have found them to be non-exponential[68–71, 206]. Even in proteins containing a single tryptophan residue, multiexponential decay kinetics have been observed suggesting mobility of the protein structure during emission[70]. However, recent time-resolved fluorescence studies of the tryptophan zwitterion and tryptophan peptides[9, 47, 48, 151, 152] indicate that this fluorophore does not decay

Table 11. Fluorescence properties of the aromatic amino acid zwitterions in aqueous solution at room temperature

λ_{max} (nm)		ϕ_f	τ_f (ns)
Tryptophan	350	0.20[a], 0.14	3.2, 3.3[b]
Tyrosine	303	0.21[a], 0.14	3.6, 3.5[c]
Phenylalanine	282	0.03	6.8

[a] Values of Teale and Weber[188]
[b] Multi exponential decay – see Table 12
[c] Rayner et al.[151]
[d] Other values from Longworth 1971[228]

Table 12. Kinetic parameters for the fluorescence decay of tryptophan derivatives

Compound	pH	$\lambda_{em} \times nm^{-1}$	τ_1/ns	τ_2/ns	R^a
Tryptophan	7	330	3.13	0.53	2.1
	7	350	3.09	0.65	4.1
	10.5	330	10.1	4.55	1.2
	1	340	0.80	0.29	2.7
	3	340	3.0	0.29	3.4
	7	340	3.3	0.43	4.3
	9.2	340	9.0	3.0	0.79
	11	340	9.1	–	∞
	13	340	1.2	–	∞
N-Acetyl-tryptophanamide	7	330	3.00	–	∞
N-Acetyltryptophan	7	330	4.80	–	∞
Tryptophanamide	5	330	1.61	–	∞
	9	330	6.86	–	∞
Tryptophan ethyl ester	5	330	1.53	0.47	0.22
	7	350	1.87	0.51	0.13
N-Acetyltryptophan ethyl ester	7	330	1.87	0.81	1.8
	7	350	1.84	0.50	2.5
5-Methyltryptophan	7	330	3.14	1.40	1.2
	7	350	2.74	0.72	4.7
6-Methyltryptophan	7	330	3.02	1.50	1.7
	7	350	2.98	1.53	1.8
Tryptophylalanine	1	340	0.97	0.24	2.1
	3	340	1.9	0.47	1.6
	7	340	5.9	1.6	0.47
	9.2	340	8.0	–	∞
	11	340	7.5	–	∞
	13	340	1.3	–	∞
Alanyltryptophan	1	340	0.58	0.17	2.8
	3	340	1.2	0.41	1.0
	7	340	2.0	0.73	1.1
	9.2	340	3.0	0.65	3.5
	11	340	2.9	0.68	4.3
	13	340	1.9	0.92	0.04
Lys Trp Lys	4.9	330	2.21	0.85	0.83
	4.9	380	2.20	0.89	0.97
Leu Trp Leu	4.9	330	2.12	0.70	0.99
	4.9	380	2.14	0.80	1.06
Glu Trp Glu	4.9	330	1.83	0.71	0.95
	4.9	380	1.86	0.77	1.03
Gly Trp Gly	4.9	330	1.64	0.71	0.60
	4.9	380	1.71	0.75	0.57
Leu Trp Met	4.9	330	2.02	0.74	1.0
	4.9	380	1.99	0.71	1.08

Table 12. (continued)

Compound	pH	$\lambda_{em} \times nm^{-1}$	τ_1/ns	τ_2/ns	R[a]
Ala Trp Ala Gly	4.9	330	3.1	1.02	0.075
	4.9	380	3.38	1.06	0.027
Glucagon	3	340	3.3	1.1	1.5
	7	340	3.7	1.1	1.8
	11	340	3.5	1.2	2.0

[a] Data compiled from Szabo and Rayner[183], Beddard et al.[9, 10]
    R is the ratio of pre-exponential factors of long component to short

by a single exponential process as suggested by previous workers. The two components
of 2.1 ns and 5.4 ns in neutral solution of aqueous tryptophan reported by Fleming
et al.[47, 48] are undoubtedly due to the accumulation of photoproducts resulting
from the use of a high-powered solid state laser for excitation. The same result can
be obtained by prolonged irradiation with a xenon arc lamp[62]. Under less intense
illumination, two components with decay times summarized in Table 12 are observed.
The two components were first attributed to emission from solvent equilibrated
^1L_a and ^1L_b states of tryptophan[151], later to separate conformational forms of the
amino acid[183]. At least in the case of the peptide glucagon (Table 12), which has 29
amino acids with tryptophan at position 25, this conformational theory has been
challenged on the grounds that at pH 11, increase in concentration from 0.2 mg/ml
to 2 mg/ml, which increases the population of trimers of the α-helical form by a factor
of 20, does not change significantly the two decay times or their relative weighting[10].
This implies the dual exponentiality is an intrinsic property of the tryptophan molecule. The interpretation of protein fluorescence is thus complicated. Further, it is
often assumed that relaxation of the fluorophore environment following light absorption occurs on a time scale much shorter than fluorescence and thus emission always
occurs from some equilibrated lowest singlet state. Weber[215] and Grinwald and
Steinberg[71] have provided evidence from quenching and decay measurements that
the relaxation of the environment in proteins may in fact occur during the decay process. Such behaviour would lead to a red-shift in emission during the fluorescence
decay as illustrated in Fig. 5. Time-resolved fluorescence spectral studies of indole
derivatives such as skatoles, XII in various solvents[118] have indicated that the rate
of relaxation is highly dependent on the microviscosity of the fluorophore environment. In viscous solvents such as octanol and lipid bilayers time dependent spectral
shifts of indole occur on a nanosecond time scale (Fig. 39). Similar results have been

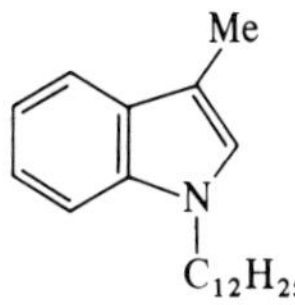

XII N-dodecyl skatole

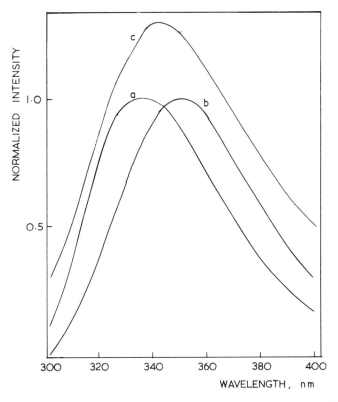

Fig. 39a–c. TRES of skatole (see text) in egg phosphatidyl choline **(a)** Early gated spectrum with $\Delta t = 0$ ns, $\delta t = 3.9$ ns **(b)** Late gated spectrum with $\Delta t = 21.3$ ns, $\delta t = 2.6$ ns **(c)** Total fluorescence spectrum

observed in dansyl derivatives XIII[64] such as dansyl amide in n-butanol at low temperatures, dansyl phosphatidyl ethanolamine in liposomes. It should be stressed that in these cases simple solvent relaxation (see Sect. A) cannot adequately explain the observed time dependent spectral shifts, and multiple siting has also to be invoked. Weber (1976) has also observed similar behaviour for the fluorescent probe 1-anilino-8-naphthalene sulphonate (ANS) XIV bound to bovine serum albumin where an energy of activation of "antirelaxation" of 15.5 kJ mol^{-1} was calculated. The possibility

XIII Dansyl amide

that such relaxation effects influence emission from the chromophores in proteins must also be seriously considered.

Although it has long been accepted that the emission from proteins in the 350 nm region is largely attributable to tryptophan, recent work by Rayner et al.[151, 152] has raised the possibility that excited state ionization of the tyrosine phenolic groups occurs in proteins under neutral pH conditions and thus tyrosinate emission may significantly contribute to protein fluorescence.

XIV 1,8 anilino
naphthalene sulphonate

It is clear from the evidence presented above that a number of factors may influence emission from proteins, these being multi-exponential decays of single tryptophan residues, heterogeneity of environment, presence of other emitting species and nanosecond fluctuations of the macromolecule structure. Only when the contributions of these phenomena have been assessed can any definite conclusions be made about the influence of structure on the photophysics of biopolymers.

The existence of electronic energy transfer between the chromophores in proteins and polynucleic acids is another question that has been discussed by authors over the years[41, 228]. For proteins there is substantial evidence from steady state luminescence data[56, 122, 228] that energy transfer occurs in the sequence phenylalanine → tyrosine → tryptophan in accordance with the observed sequence of singlet energies. Evidence for intertryptophan energy migration has also been obtained from fluorescence polarization data[111, 122]. In the case of polynucleotides and various DNA's several studies have given some insight into how the excitation energy initially located in pyrimidine and purine bases migrates to thymine residues, although observations of luminescence from these biopolymers are complicated by the low quantum yields at room temperature[41, 80]. However, the dynamic properties of biomolecules may have considerable effects on energy transfer processes. Solvent relaxation effects which will alter the donor-acceptor spectral overlap requirements during emission and the changes in orientation of the interacting chromophores during the excited state lifetime could profoundly influence the rates of energy transfer. Observations of the grow-in and decay of donor and acceptor emission and/or measurements of the time dependence of emission anisotropy promise to provide much more insight into energy migration processes although to date only in certain model compounds and photosynthetic systems have such observations been made using picosecond laser techniques[160, 161]. The application of time-resolved luminescence methods to observe directly energy migration mechanisms in biologically important macromolecules should resolve many of the uncertainties in this important area of research.

III. Applications of Fluorescence Probe Techniques

The binding of various fluorescent probe molecules to biological systems is often accompanied by changes in decay time and emission characteristics of the probe[25, 75, 181]. These changes have been used to provide information about the environment of the fluorescent probe and to follow changes in conformation of the macromolecule. In other work the study of the fluorescence polarization properties of the attached probe under steady state illumination and the application of Perrin's equation[25] enable calculation of the rotary Brownian motion of the polymer. This technique has been extended by Jablonski[96] and Wahl[201] to the use of time-resolved fluorescence polarization measurements to calculate rotational relaxation times of molecules[187]. These experiments are discussed fully in the following section of this review.

Considering the wide-spread use of fluorescent probe techniques in biological studies, it is important that a full understanding of the photophysics of the bound and unbound probe is available to enable unambiguous interpretations of the fluorescence data. The most common probes used for fluorescence work are the anilino-naphthalene sulphonate derivatives, particularly 1-anilino-8-naphthalene sulphonate (ANS) (XIV) and 2-p-toluidinyl-6-naphthalene sulphonate (TNS) (XV). Changes in fluores-

XV TNS

cence yields, absorption profiles and fluorescence spectra accompanying the binding of these derivatives to biomolecules from aqueous solution have been used as a measure of the polarity of the binding site[40]. However, a number of recent studies have suggested that solvent cohesion and structure[153], the nature of the electronic states involved[112, 168] and environment rigidity[18, 33] may also play an important role in determining the photophysical properties of these molecules. Picosecond laser studies have also emphasised the importance of solvent effects on electron photojection of anilino-sulphonate derivatives[44, 160, 161].

In our own studies we have investigated the binding of ANS and TNS in lipids, proteins and other cell constituents using time-resolved fluorescence methods[118, 120]. TNS bound to liposomes of egg yolk phosphatidylcholine gives rise to complex fluorescence decay kinetics and the lifetime increases with increasing observation wavelength (Fig. 40). Time-resolved emission spectra of this system (Fig. 41) show a gradual change from a slightly structured blue-shifted emission at early time-gates to a broad structureless red-shifted emission at late observation times. These observations are similar to those of Weber[215] for ANS bound to bovine serum albumin and the reported behaviour of TNS in several biochemical systems[38, 39] and are consistent with excited-state solvent relaxation processes occurring on a nanosecond time scale.

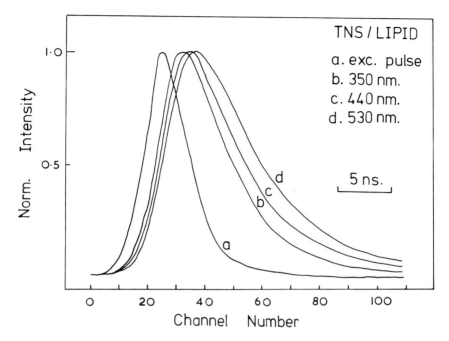

Fig. 40. Fluorescence decay curves recorded at different emission wavelengths for TNS bound to liposomes of egg yolk phosphatidylcholine at 20 °C

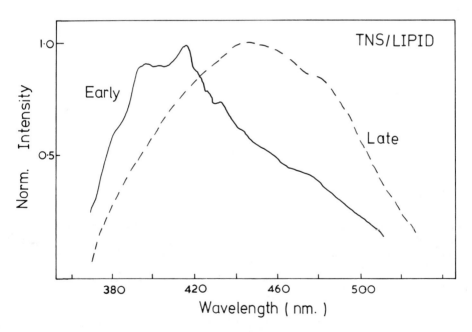

Fig. 41. Time-resolved emission spectra for TNS bound to egg lipid at 20 °C. Early spectrum: time delay (Δt) = 0 gate width (δt) = 2.5 ns. Late spectrum: Δt = 31 ns, δt = 2.5 ns

Obviously the dipole characteristics of the fluorescent probe and the solvent together
with the rigidity of the solvent environment will determine the possibility and extent
of any reorientation process. The sensitivity of these fluorescent probes to environmental relaxation processes emphasise that care is required in interpreting the fluorescence behaviour in biochemical systems purely in terms of the polarity of the bonding site. The possibility that intramolecular charge-transfer may occur in the excited
states of ANS[113] or that specific solute-solvent complexes may form[48] further com-
plicate the photophysics of these molecules.

   Although the above discussion will have illustrated the difficulties in the application of fluorescent probe techniques to biopolymers, it has been shown that under
certain conditions[118, 120] the different spectral and temporal behaviour of ANS
bound to lipids and proteins may be used to investigate qualitatively the effect of
drug binding in a heterogeneous biological membrane. In these experiments the emission decay from ANS bound to membrane samples has been isolated into components
arising from lipid and protein sites. The increase in the relative amount of long-lived
component (ascribed to ANS emission from protein environments) upon addition of
barbiturate drug combined with the decrease in overall fluorescence intensity may be
interpreted as evidence for the drug displacing ANS from its lipid binding site
(Fig. 42). Further time-resolved fluorescence studies are required in this area to test
the general application of this technique, which at present shows promise.

   It is clear that in all cases mentioned above, time-resolved fluorescence methods
have enabled a far greater insight into the character and complexities of biopolymer
luminescence than have previously been obtained by steady state illumination techniques.

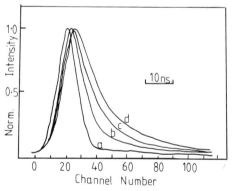

Fig. 42. Fluorescence decay curves for ANS (1.7×10^{-5} M) bound to microsomal membranes
(1 ng/ml) from *tetrahymena pyroformis* recorded at 480 nm in the presence of pentobarbitone
(a) excitation pulse (b) without pentobarbitone (c) with 10^{-6} M pentobarbitone and (d) with
5×10^{-6} M pentobarbitone. After Lee et al.[118]

E. Time-Resolved Fluorescence Depolarization in Macromolecules

I. Introduction

Time-resolved fluorescence depolarization studies have, over the past decade, provided an interesting method for monitoring molecular reorientational motions in solution. The technique has been applied to investigations of both synthetic polymers and macromolecules of biological interest, and a selection of the results of these are discussed here. However, until recently, the relatively poor quality of much of the data available from these measurements has precluded detailed quantitative interpretations of the results. With the advent of improved experimental techniques for fluorescence decay time determinations due in part to the availability of pulsed lasers for sample excitation and more accurate data analysis procedures, it is envisaged that interest in the technique may be revived. We will present here a short recapitulation of the background to these experiments, following from Sect. A. V.

If an ensemble of fluorescent chromophores are illuminated by a pulse of polarised light then an anisotropy will be created in the system since only those species with a component of the absorption transition moment in the same direction as the polarization vector of the incident light will be excited. If the fluorescence emission transition moment maintains a fixed geometrical relationship with that for the absorption process, then as stated earlier three distinct cases may be encountered. Firstly, if the fluorescence decay time, τ_f is considerably shorter than the rotational relaxation time, τ_r, then the anisotropy in the system will be unchanged during the excited state lifetime, and the resultant fluorescence will be polarized. Alternatively, if the molecular reorientation is fast compared to the fluorescence time scale, then the anisotropy will rapidly decay to a limiting value. For an isotropic system such as a chromophore rotating in a small molecule, low viscosity solvent, this value will be zero and the fluorescence will be completely depolarised. For some systems, however, such as a chromophore rotating in an ordered solvent such as a lipid bilayer or liquid crystal, the depolarization may not be total, indicating a measure of anisotropy fundamental to the system. The third case lies between the first two and occurs when the fluorescence and molecular reorientation occur on a similar timescale ($\tau_f \simeq \tau_r$). In such a system, the time course of the decay of a polarised component of fluorescence will be modulated due to the molecular motion. The time-resolved fluorescence depolarization experiment utilises this fact to derive the time course of the rotational correlation function for the reorientational motion.

Although the relation between fluorescence depolarisation and rotational Brownian motion was first identified by Perrin[142, 143] and the development of the theoretical background of the time-resolved fluorescence depolarization experiments was made by Jablonski[96−98], use of the technique was limited until the advent of improved fluorescence decay time measurements some fifteen years ago. An alternative, related technique, involving excitation using a continuous polarised light source, provides only the time average of the correlation function (Eq. 18) and as such, is less useful than the time-resolved method. Other disadvantages are that the natural decay time of the chromophore must be determined from a separate experiment and it is necessary to alter the viscosity, and/or temperature of the medium, often with unknown effect on the conformation of the macromolecule. The application of this "steady state" technique to macromolecular systems has been recently reviewed[4] and will not be discussed in detail here.

The usual experimental technique is to observe the fluorescence of a probe molecule incorporated into the system under investigation and make the assumption that the motion of the probe reflects that of its environment. Probe molecules may be divided into two categories: intrinsic probes (such as fluorescent amino acid residues naturally occurring in a protein under investigation) and extrinsic probes either bound chemically to the macromolecule or simply dispersed in the system. The criteria that need to be applied when selecting suitable candidates for probe molecules have been discussed by Stryer[181]. Ideally the probe should

(i) specifically label one region of the macromolecule,

(ii) not perturb the system under investigation (in the case of biologically important macromolecules the physiological function of the system must also remain unchanged[189]), and

(iii) have well quantified spectral properties suitable for the measurement required.

For time-resolved fluorescence depolarization studies, this last consideration demands that the probe have a simple fluorescence decay law in isotropic systems and that the fluorescence decay occurs at a similar rate to the reorientational motion.

The major disadvantage with probe techniques lies in the assumption that the rotational motion of the probe is a good indicator of that of the host system. In the light of this it is important to discuss the relative merits of the fluorescence depolarization experiment over other techniques such as dielectric relaxation[185], flow birefringence[186], electric birefringence, electric dichroism[140], quasi-elastic light scattering[22, 141], Raman Spectroscopy[165], nuclear magnetic resonance[37], conventional electron paramagnetic resonance[182] and saturation transfer epr[189]. The effective time ranges of some of the more important of these methods are shown diagramatically in Fig. 43. As may be seen, fluorescence depolarisation measurements permit

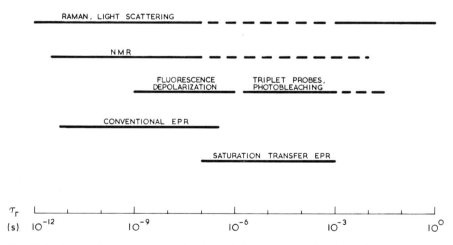

Fig. 43. Techniques for measuring rotational correlation times (τ_r) of proteins, showing the approximate time range over which each method has proven effective. (After D. D. Thomas[189])

rotational relaxation times, τ_r, to be measured in the range 1 ns to 1 μs corresponding to the rotation in aqueous solution, of species with sizes up to 1 million Daltons[209]. The range can be extended down to around 10 ps using pulsed laser excitation, and to the ms region by using phosphorescent probes. An advantages of the techniques involving motion sensing spectroscopic probes is that selected regions of the macromolecule may be labelled thus permitting observation of internal motions of the polymer in addition to the total molecular rotation. Of the probing techniques, fluorescence depolarization has the advantage of a wide selection of suitable probe molecules; epr methods usually being confined to nitroxide spin labels. Another advantage of the fluorescence technique is that the time profile of the 2nd order correlation function for the rotational motion is directly evaluated. Time-correlation functions may be predicted from theories of molecular motion, and from computer simulation studies, and so the fluorescence depolarization technique provides a means of testing these experimentally. A review of the use of correlation function and their relationship to molecular rotation and experimentally determined quantities has been published[220]. For ordered macromolecular systems, such as liquid crystals, the fluorescence depolarization measurement provides one of the few experimental techniques capable of measuring 4th rank order parameters[99, 225]. These are of central importance in any theory of the liquid crystalline state.

The applications of the technique may be divided into two categories. In one the motion of the whole or part of the macromolecule may be measured, providing information concerning the size and shape of the species, internal flexibility, aggregation states, etc. In the other case the environment surrounding the macromolecule may be probed providing information concerning, for example, the microviscosity of lipid bilayers. Examples of both cases will be presented here.

The experimental techniques used for these measurements are similar to those discussed in the Sect. B for fluorescence decay curve determinations and only those points particularly pertinent to time-resolved fluorescence depolarization will be considered here. The sample is excited using a short pulse of monochromatic, plane polarized light (1–2 ns wide) from either a nanosecond spark discharge lamp or a pulsed laser. The fluorescence is monitored at right angles to the excitation light path, through an analysing polarizer which may be set to transmit fluorescence polarized either parallel to or perpendicular to the polarization of the excitation. Conventional detection techniques are employed. Special problems associated with time-resolved fluorescence depolarization measurements, such as the finite collection angle, have been discussed by Zinzli[227] and Tschanz and Binkert[191]. The experimental geometry is shown in Fig. 44. The sample is placed at the origin and $\vec{\mu}$ is considered to be a transition dipole fixed in the body. For simplicity the special case in which the emission and absorption transition dipoles are parallel will be considered initially; the extension to the general case will be discussed later. The sample is excited with light incident along the x-axis and plane polarized parallel to z. The fluorescence is monitored along the y-axis. The two analysing polarizer positions, $I_\parallel$ and $I_\perp$, are shown, and correspond to the detection of fluorescence polarized respectively parallel and perpendicular to the exciting light. The emission anisotropy, r, defined in Eq. 14, is a more useful parameter than the degree of polarization (Eq. 13), in that many of the expressions related to polarization of photoluminescence are much simplified[96–98] and r may be directly related to the correlation function for molecular

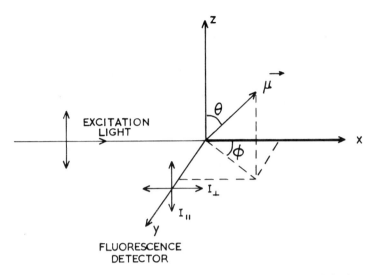

Fig. 44. Experimental geometry for time-resolved fluorescence depolarization measurements

reorientation. A lucid description of this relationship has been given by Tao[187], a
brief review of which is presented here. For the time-resolved fluorescence depolarization experiment, the time dependence of the emission anisotropy, r(t), is required as
defined in Eq. 14.

    With the experimental geometry shown in Fig. 44, the following quantities may
be defined: $W(\theta, \phi, t)$ is the probability that the dipole $\vec{\mu}$ will be oriented at angle
(θ, ϕ) at time t, $G(\theta_0, \phi_0/\theta, \phi, t)$ is the probability that $\vec{\mu}$ will be oriented along
(θ, ϕ) at time t given that it was oriented along (θ_0, ϕ_0) at time t = 0, and $W(\theta_0, \phi_0)$
is the distribution of the orientations of $\vec{\mu}$ at t = 0. Thus

$$W(\theta, \phi, t) = \int_0^{2\pi} d\phi_0 \int_0^{\pi} \sin\theta_0 \, d\theta_0 W(\theta_0, \phi_0) G(\theta_0 \, \phi_0 | \theta \phi t) \tag{83}$$

The probability of a molecule absorbing a photon is directly proportional to the second power of the component of the transition moment in the same direction as the
polarization of the exciting light. Thus, for pulsed sample excitation at time t = 0

$$W(\theta_0 \, \phi_0) = (3/4 \, \pi) \cos^2 \theta_0$$
$$= (1/4 \, \pi) [1 + 2P_2(\cos \theta_0)] \tag{84}$$

where $P_2(\cos \theta_0)$ is the second order Legendre polynomial. $G(\theta_0 \, \phi | \theta \phi t)$ is now ex-
panded in terms of surface harmonics:

$$G(\theta_0 \, \phi_0 | \theta \phi t) = \sum_{l=0}^{\infty} \sum_{m=-1}^{1} C_{l,m}(t) \, Y^*_{l,m}(\theta_0 \, \phi_0) \, Y_{l,m}(\theta \phi) \tag{85}$$

where $C_{1,m}(t)$ are the expansion coefficients. With appropriate boundary and normalization conditions, consideration of the above expressions leads to

$$W(\theta, \phi, t) = (1/4\,\pi)\,[1 + 2C_{2,0}(t)\,P_2(\cos\theta)] \tag{86}$$

The intensity of the fluorescence at time t, polarized at a particular angle to the excitation polarization, is proportional to the second power of the component of the emission dipole in the same direction and the probability $P(t)$ that the sample is still excited at time t. For an individual transition dipole, the intensity of light polarized parallel and perpendicular to the exciting light ($J_{\parallel}(t)$ and $J_{\perp}(t)$) at time t will be given by

$$J_{\parallel}(\theta, \phi, t) \alpha\, P(t)\, \mu^2\, \cos^2\theta \tag{87}$$

$$J_{\perp}(\theta, \phi, t) \alpha\, P(t)\, \mu^2\, \sin^2\theta\, \cos^2\phi \tag{88}$$

For the total fluorescence intensity at time t the above expressions are averaged over all orientations. Thus

$$I_{\parallel}(t) = \int\!\!\int\, \sin\theta\, d\theta\, d\phi\, J_{\parallel}(\theta, \phi, t)\, W(\theta, \phi, t) \tag{89}$$

$$I_{\perp}(t) = \int\!\!\int\, \sin\theta\, d\theta\, d\phi\, J_{\perp}(\theta, \phi, t)\, W(\theta, \phi, t) \tag{90}$$

From the expression derived earlier it follows that

$$I_{\parallel}(t) = \left[\frac{1}{3} + \frac{4}{15} C_{2,0}(t)\right] P(t) \tag{91}$$

$$I_{\perp}(t) = \left[\frac{1}{3} - \frac{2}{15} C_{2,0}(t)\right] P(t) \tag{92}$$

from which may be derived

$$r(t) = \frac{2}{5} C_{2,0}(t) \tag{93}$$

It should be noted that the decay curve measured in the absence of an emission polarizer will be $I_{\parallel}(t) + I_{\perp}(t)$ whereas inspection of the above demonstrates that the true fluorescence decay curve will be proportional to $I_{\parallel}(t) + 2I_{\perp}(t)$. Thus, for molecules rotating on the same timescale as the fluorescence decay, the fluorescence decay curve measured in the absence of a polarizer, will be distorted, and so time-resolved fluorescence depolarization must be considered even when the only desired measurement is the intrinsic fluorescence decay of the chromophore. To overcome this problem the fluorescence should be monitored through a polarizer set at 54.7° to the excitation polarization vector[43].

    The correlation function for molecular reorientation $\langle P_2[\hat{\mu}(o) \cdot \hat{\mu}(t)]\rangle$ where $\hat{\mu}(t)$ is a unit vector fixed in the rotating body at orientation (θ, ϕ) at time t is given by

$\langle P_2[\hat{\mu}(o) \cdot \hat{\mu}(t)]\rangle = 1/4 \pi \iint \sin\theta \, d\theta \, d\phi \iint \sin\theta_0 \, d\theta_0 \, d\phi_0 .$

$P_2[\hat{\mu}(o) \cdot \hat{\mu}(t)] \cdot W(\theta_0 \phi_0) \, G(\theta_0 \phi_0 | \theta \phi t) = C_{2,0}(t)$

$$\text{(94)}$$

and so

$$r(t) = \frac{2}{5} \langle P_2[\hat{\mu}(o) \cdot \hat{\mu}(t)]\rangle \tag{95}$$

This result has also been derived by Gordon[67]. The extension to the general case, with emission and absorption transition dipoles at angle α, gives the result

$$r(t) = \frac{2}{5} P_2(\cos\alpha) \, C_{2,0}(t) \tag{96}$$

Thus, as mentioned earlier, time-resolved depolarization measurements afford a means of recording the time profile of the rotational autocorrelation function. The steady state technique, with continuous sample excitation, produces merely the time average of the emission anisotropy, $\bar{r}$. For a rotating chromophore with a single fluorescence decay time τ_f, $\bar{r}$ is related to r(t) by the following expression

$$\bar{r} = \frac{1}{\tau_f} \int_0^\infty r(t) \, e^{-t/\tau_f} \, dt \tag{97}$$

a result which may be used to check the experimental data from the time-resolved measurement.

Several attempts have been made to derive the form for the emission anisotropy for a general, asymmetric rotating body. Lombardi and Dafforn[123] and Weber[214] derive expressions in which r(t) is described by two exponentially decaying terms. Tao[187] on the other hand, finds an expression in which five exponential terms are required. In addition to this discrepancy, Lombardi and Dafforn, Weber, and Memming[125] predict the possibility of an increase in r(t) initially whilst Tao's treatment does not allow for this. The correct form for the emission anisotropy for a general rotating body was derived by Belford et al.[11] and was found to contain five exponential terms:

$$\frac{5}{6} r(t) = \sum_{i=1}^{3} C_i \exp(-t/\tau_i) + [(F + G)/4] \exp(-6D - 2\Delta \ t)$$

$$+ [(F - G)/4] \exp(-[6D + 2\Delta]t) \tag{98}$$

where,
$D = (D_1 + D_2 + D_3)/3$, the mean rotational diffusion constant given that D_1, D_2 and D_3 are the rotational diffusion constants around the three principal axes labelled 1, 2 and 3.

$$\Delta = (D_1^2 + D_2^2 + D_3^2 - D_1 D_2 - D_1 D_3 - D_2 D_3)^{1/2}$$

$$C_i = \alpha_j \alpha_k \epsilon_j \epsilon_k \quad \text{where (ijk)} = (123), (231) \text{ or } (312)$$

$\alpha_1, \alpha_2, \alpha_3$ are the direction cosines of the absorption dipole with respect to the principal axes.

ϵ_1, ϵ_2 and ϵ_3 are, likewise, the direction cosines of the emission dipole.

$$\tau_i = (3D + 3D_i)^{-1}$$

$$F = \sum_{i=1}^{3} \alpha_i^2 \epsilon_i^2 - 1/3$$

$$G\Delta = \sum_{i=1}^{3} D_i(\alpha_i^2 \epsilon_i^2 + \alpha_j^2 \epsilon_j^2 + \alpha_k^2 \epsilon_k^2) - D \quad i \neq j \neq k \neq i$$

The expressions derived by Lombardi and Dafforn[123] and Weber[214] are greatly simplified since special symmetry of either the absorption or emission dipole was implicit in the treatments. The derivation by Tao[187] correctly predicts five exponential terms but disagrees in the form of the pre-exponential factors for the general case in which the absorption and emission dipoles are not parallel since he assumes a one-to-one correspondence between the distribution of body orientations and the distribution of orientations of the emitting dipole[11]. Computer simulated data have been used to check the validity of the general expression[77]. In all cases good agreement was observed between the simulated data and theoretical curves. The results for one such experiment are shown in Fig. 45 in which it is interesting to note the initial increase in r(t) as a consequence of the orientation of the absorption and emission dipoles.

From the above it is apparent that the time profile of the emission anisotropy is very complex for a general, asymmetric rotator. Thus it is unlikely that, even with the superior data obtained with pulsed lasers for sample excitation, meaningful analy-

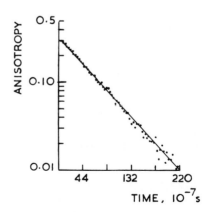

Fig. 45. Computer simulated emission anisotropy (*points*), compared with theoretically predicted r(t) from the equation by Belford et al. (1972) (*line*). (After Harvey and Cheung[77])

sis in terms of five exponentials will be possible, and experiments are limited to studies of the simpler systems discussed below. In connection with this it is interesting to note that Small and Isenberg[172] have suggested that any measured emission anisotropy will be described well by, at most, three exponential terms since several of the decay times in the expression for r(t) above are similar in magnitude.

Several simplifying cases may now be considered. In general, for excitation into the lowest singlet state, the absorption and emission dipoles are parallel. This condition will be assumed in the following discussion. For a typical experiment in which a macromolecule is labelled with a fluorescent probe attached to random sites, the expression for r(t) is reduced to

$$r(t) = \frac{2}{15} \left[\exp\left(-3(D_1 + D)t\right) + \exp\left(-3(D_2 + D)t\right) + \exp\left(-3(D_3 + D)t\right) \right] \quad (99)$$

For a body that possesses an axis of symmetry, such as an ellipsoid of rotation, we may define

$$D_1 = D_2 = D_\perp$$

$$D_3 = D_\parallel$$

The emission anisotropy in this case is given by[187]:

$$r(t) = \frac{2}{5} \left[A_1(\theta) \exp(-6D_\perp t) + A_2(\theta) \exp(-(5D_\perp + D_\parallel)t) \right.$$

$$\left. + A_3(\theta) \exp(2D_\perp + 4D_\parallel \quad \text{where} \right. \qquad (100)$$

$$A_1(\theta) = (3/2 \cos^2\theta - 1/2)^2$$

$$A_2(\theta) = 3 \cos^2\theta \sin^2\theta$$

$$A_3(\theta) = 3/4 \sin^4\theta$$

and θ is the angle between the emission (and absorption) transition dipole and the axis of symmetry. For the special case in which θ is $90°$ (e.g. the dipole is oriented perpendicular to the axis of symmetry)

$$r(t) = \frac{1}{10} \exp(-6D_\perp t) + 3 \exp(-(2D_\perp + 4D_\parallel)t) \qquad (101)$$

and for the dipole parallel to the axis ($\theta = 0$) the equation for r(t) is reduced to a single exponential term:

$$r(t) = \frac{2}{5} \exp(-6D_\perp t) \qquad (102)$$

For a rotating spherical body, the correlation function of order n is given by[220]

$$\langle P_n[\hat{\mu}(0) \cdot \hat{\mu}(t)]\rangle = \exp[-n(n+1)D_r t] \tag{103}$$

Thus, for fluorescence depolarization measurements in which the second order correlation function is desired

$$r(t) = \frac{2}{5} \exp(-6D_r t) \tag{104}$$

Here D_r is the rotational diffusion constant and is given by the Einstein law as

$$6D = \frac{1}{\tau_r} = \frac{kT}{V\eta} \tag{105}$$

where V is the volume of the sphere and η is the viscosity of the solvent. Some confusion exists in the definition of the rotational relaxation time τ_r, many reports preferring $\rho_0 (= 3\,\tau_r)$.

Table 13. Rotational diffusion decay constants for prolate and oblate ellipsoids

ρ	$1/\rho$	$\tau_\parallel/\tau_r$	τ_1/τ_r	τ_2/τ_r	τ_3/τ_r
Prolate ellipsoids					
1		1.0000	1.0000	1.0000	1.0000
2		0.8067	1.5049	1.3152	0.9543
3		0.7480	2.3408	1.7276	0.9674
4		0.7210	3.3956	2.0984	0.9777
5		0.7061	4.6405	2.4060	0.9842
6		0.6968	6.0616	2.6548	0.9884
7		0.6906	7.6506	2.8549	0.9911
8		0.6862	9.4008	3.0162	0.9930
9		0.6829	11.3080	3.1472	0.9944
10		0.6805	13.3680	3.2545	0.9954
15		0.6739	25.8607	3.5774	0.9979
20		0.6712	41.8199	3.7280	0.9988
Oblate ellipsoids					
1		1.0000	1.0000	1.0000	1.0000
2		1.4100	1.1316	1.1701	1.3032
3		1.8284	1.4645	1.5147	1.6885
4		2.2495	1.8431	1.9003	2.0955
5		2.6718	2.2398	2.3018	2.5104
6		3.0948	2.6455	2.7111	2.9290
7		3.5181	3.0564	3.1247	3.3495
8		3.9418	3.4705	3.5411	3.7711
9		4.3655	3.8869	3.9593	4.1934
10		4.7894	4.3049	4.3787	4.6162
15		6.9099	6.4073	6.4859	6.7338
20		9.0312	8.5193	8.6005	8.8538

ρ is the axial ratio (ratio of longitudinal semiaxis to equatorial semiaxis) $\tau_\parallel = (6D_\parallel)^{-1}$, $\tau_1 = (6D_\perp)^{-1}$, $\tau_2 = (5D_\perp + D_\parallel)^{-1}$, $\tau_3 = (2D_\perp + 4D_\parallel)^{-1}$, $\tau_r = (6D_r)^{-1}$ is the relaxation time under the spherical model. Adapted from Tao [187]

The three decay constants appearing in the expression for r(t) for an ellipsoid of rotation have been calculated and are shown in Table 13. As may be seen, the three relaxation times diverge rapidly with increasing axial ratio for a prolate ellipsoid. However, for an oblate ellipsoid the deviations are small even for high axial ratios and experimentally it may prove difficult to resolve more than a single, mean relaxation time for r(t) in this case. Thus, three situations exist in which the emission anisotropy may decay exponentially and it is not possible, therefore, to distinguish
(i) a true spherical geometry,
(ii) oblate geometry,
(iii) preferential orientation of the fluorescence probe with respect to the principal axis.

Hence, it should be stressed, a single exponentially decaying emission anisotropy does *not* uniquely correspond to a spherical rotating body. On the other hand, the observation of non-exponential behaviour indicates deviations from spherical behaviour.

The form for the emission anisotropy in several special cases has been derived. The decay of the anisotropy for a chromophore rotating between two barriers has been derived by Wahl[208]. The problem posed by a fluorophore with restricted rotation in an ordered lipid bilayer has been discussed by Kinosita et al.[103]. If r_0 is the initial anisotropy created in the system at time 0 and r_∞ is the final limiting value after long times, then, unlike an isotropic, low viscosity solvent, r_∞/r_0 will not be zero, but will be a measure of the degree of confinement of the orientation of the label imposed by the architecture of the membrane. The decay in r(t) between these two limits should reflect the degree of "wobbling" motions possible for the fluorophore.

The segmental motions in a polymer chain have been considered[194-196]. With the chain undergoing solely three-bond motions on a tetrahedral lattice, the emission anisotropy will be given by

$$r(t) = r_0 \exp(t/\rho)\operatorname{erfc}\sqrt{\frac{t}{\rho}} \tag{106}$$

where erfc is the error function complement and ρ is the relaxation time which may contain parameters descriptive of the diffusion jump frequency and the conformational structure of the chain. When both valence angles and internal rotation angles are allowed to vary from those associated with an ideal lattice, an additional exponential term is introduced[34] and the expression for r(t) becomes

$$r(t) = r_0 \exp(-t/\theta) \exp(t/\rho)\operatorname{erfc}\sqrt{\frac{t}{\rho}} \tag{107}$$

where θ is a relaxation time reflecting the relaxation with respect to the ideal lattice.

In general, therefore, the time dependence of the emission anisotropy will be complex. Because of this, a high signal-to-noise ratio is desirable for r(t) so that meaningful data analysis may be performed. To achieve this it is necessary to accumulate $I_\parallel(t)$ and $I_\perp(t)$ to a very high signal-to-noise ratio[89] ($>10^5$ counts in the maximum) since the anisotropy, r(t), is derived from the difference between these two, similar functions. Thus the advantage of fast data acquisition times achieved with pulsed lasers for sample excitation becomes apparent. An additional problem is that, for decays occurring on a similar time-scale to the excitation pulse width, the recorded decay curves will be distorted and the true decay profile can only be obtained with the use of a deconvolution procedure[136]. The inaccuracies in the deconvolution analysis are reduced with the high quality data available with pulsed laser excitation and so, in this case, rotational correlation times which are short with respect to the excitation pulse width, may be measured with confidence. To date, however, the use of laser excitation for time-resolved fluorescence depolarization studies has been largely confined to small molecules rotating on the sub-nanosecond timescale (see for example[43–46, 149, 160, 161]).

The alternative method for fluorescence decay time determinations, discussed in the measurement of the phase difference and depth of modulation of the fluorescence, has been shown to be also suitable for the measurement of rotational correlation times[178]. The technique has been little used, however, but is mentioned in passing since it has the potential for high temporal resolution. The theory developed for simple, spherical rotors has been extended to include rotating irregular molecules[217]. One application involving macromolecules is the investigation of the changed rotational motion of the fluorescent probe 1-phenylnaphthylamine bound to cells from Escherichia coli on the deenergization of the membrane with the addition of colicin E1[215, 216]. Other applications of this technique will be presented later, at the appropriate point, in the discussion of the results from the more widely used pulse fluorometry method.

II. Applications

1. Rotational Correlation Times of Proteins

An early application of time-resolved fluorescence depolarization measurements was in the determination of the rotational correlation times of globular proteins in solution. In one such experiment[202] Bovine Serum Albumin was labelled with dimethyl amino 1-sulphonyl 5-naphthalene DNS and dissolved in aqueous solution at pH 7.2. The sum $S(t) = (I_{\parallel}(t) + 2I_{\perp}(t))$ and difference $D(t) = (I_{\parallel}(t) - I_{\perp}(t))$ were derived from separate measurements of $I_{\parallel}(t)$ and $I_{\perp}(t)$ as discussed above. These are shown in Fig. 46. As may be seen, both are well described, over two decades of intensity, by a single exponential decay law and so one relaxation time, τ_r, only may be derived from the emission anisotropy. This is shown, along with results for other globular proteins, in Table 14. The full potential of time-resolved fluorescence depolarization measurements is not realised in these experiments, however, since the motion of the whole body has been investigated. Indeed, this application of the technique has fallen into a decline in popularity since the same information may be obtained from several of the alternative techniques, listed at the beginning of this section without the disadvantages associated with the use of spectroscopic probes.

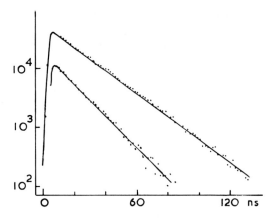

Fig. 46. Sum S(t) and difference D(t) curves for Bovine Serum Albumin labeled with DNS in aqueous solution at pH 7.2 S(t) – *upper curve*, D(t) – *lower curve*. *Points* are experimental data, *lines* are best single exponential fit to data. (After Wahl[202])

Table 14. Rotational correlation times of proteins determined by nanosecond polarization spectroscopy

Protein	Molecular weight	Observed τ_r[a]/ ns	τ_r obs[b]/ τ_r calc	Ref.
Apomyoglobin	17 000	8.3	1.9	d
β-Lactoglobulin	18 400	8.5	1.8	e
Trypsin	25 000	12.9	2.0	c
Chymotrypsin	25 000	15.1	2.3	f
Carbonic anhydrase (dimer)	30 000	11.2	1.4	c
β-Lactoglobulin	36 000	20.3	2.1	e
Apoperoxidase	40 000	25.2	2.4	d
Serum Albumin	66 000	41.7	2.4	g

[a] Adjusted for temperature and viscosity of water at 25 °C
[b] τ_r calc is calculated for a rigid, unhydrated sphere of the same molecular weight as the protein, assuming a partial specific voulme of 0.73 ml/g
[c] Yguerabide et al.[222]
[d] Tao[187]
[e] Wahl and Timasheff[203]
[f] Stryer[181]
[g] Wahl[202]
    From Yguerabide[223]

2. Segmental Motions in Synthetic Polymers

In an early report, North and Soutar[135] investigated several polymer systems with fluorescent labels situated at the chain end or interior, using both steady-state and time-resolved fluorescence depolarization measurements. Results were derived to show that rotational relaxation times for groups fixed at the chain ends were less than one fifth of those for the same groups at the centre of the polymer, reflecting the constraints to molecular rotation imposed by the chain. Although in many cases the experimental errors for the rotational correlation times were high ($\sim$80%), due to the adoption of a pulse sampling method rather than single-photon counting for decay time measurements, the non-steady state technique was necessarily better suited for the characterisation of the motion of groups in stiff chains where rotation is markedly anisotropic.

Experiments have been performed[194, 197, 198] with the object of testing the form for the autocorrelation function for segmental motion in polymers discussed earlier (Eq. 107). Polystyrene was labelled with anthracene units linked across the 9, 10 position. For the lowest excited singlet state, the absorption and emission dipole moments are also oriented across the 9, 10 positions and thus the dipole observed in the time-resolved fluorescence depolarization experiment is oriented parallel to the chain. The decays of the emission anisotropy, determined at several viscosities are shown in Fig. 47. The experimental uncertainties are indicated by the shaded region. The curvature in the plots clearly indicates that a single exponential decay function is insufficient to describe the behaviour. Satisfactory fits to functions of the type given by Eq. 107 are obtained, however, and the functions giving best agreement are also shown in the figure. Discrepancies may be observed at early times, and these may be due to

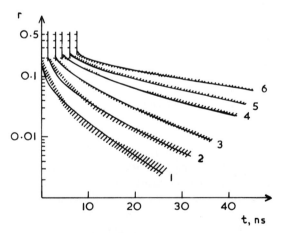

Fig. 47. Decay of fluorescence anisotropy for polystyrene labelled with anthracene in various viscosities. Theoretical curves (*drawn as bold lines*) from Eq. 107. For clarity each curve has been displaced by 1.5 ns with respect to the preceding one.

Curve	1	2	3	4	5	6
Viscosity/cp	0.43	0.89	1.50	2.88	4.35	7.47

(After Valeur and Monnerie[196])

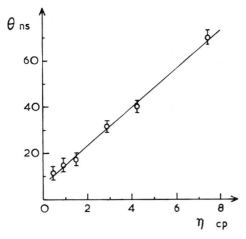

Fig. 48. Plot of relaxation time θ versus viscosity for polystyrene labeled with Anthracene groups. (After Valeur and Monnerie[197])

(i)   approximations used in the derivation of the functional form for r(t)
(ii) perturbations of the chain dynamics due to incorporation of the anthracene probe molecule, and
(iii) errors introduced due to the response of the detection system at early times.

The relaxation parameter, θ, derived from these fits displays a linear variation with solvent viscosity, as shown in Fig. 48. θ, therefore, may be expressed by a sum of two terms

$$\theta = a + b\eta \tag{108}$$

where η is the viscosity of the solvent. The parameter a depends solely upon the chemical structure of the chain and reflects either variations in the bond angles from the tetrahedral angle ($109°\ 28'$) or variations in the angle of rotation away from $120°$. The parameter b probably depends upon the variation, with viscosity, of the equilibrium of trans and gauche configurations in the polymer, the model being based upon motions on a diamond lattice[197].

Polystyrene labelled such that the emission dipole was perpendicular to the chain was also investigated using 9,10 diphenyl anthracene probe molecules[197]. As expected, this dipole orientation was more sensitive to local relaxations of the polymer chain and this was reflected in a faster decay of the emission anisotropy in this case. The polystyrene/9,10 diphenylanthracene polymer system has also been investigated by Wahl et al.[204] who found that the emission anisotropy could be well described by a model for the polymer chain resembling a "necklace" of rotating ellipsoids. The validity of the correlation function (Eq. 107) has also been confirmed using steady-state depolarization and fluorescence quenching measurements[198]. A full test of the theoretical derivations of the rotational correlation function has not been possible, however, largely due to the relatively poor quality of the experimental data, especially at early times after excitation.

3. Flexibility in Biopolymers

The time resolved fluorescence depolarization technique provides a useful method for monitoring flexible motions occurring in macromolecules since, with careful selection of the fluorescent probe molecule, a specific region of the polymer may be labelled. One system which has been extensively studied in this respect is the antibody Immuno-globulin G (IgG). The structure proposed for this molecule, based upon results from electron microscopy[193] is shown in Fig. 49. The molecule has, essentially, a "Y" shape with three segments; the two antigen binding fragments (F_{ab}) joined to the third (F_c) by a flexible hinge. The extent of this flexibility has been proposed to be important in the formation of antibody-antigen complexes[42, 134]. To determine the extent of flexibility about the hinge, rabbit immunoglobulin G was labelled with a fluorescent hapten (dansyl-Lys; ϵ-1-dimethylamino-5-naphthalene sulphonyl-L-lysine) attached at the active sites of the antibody (ends of the F_{ab} sections)[222]. In additional to this, two smaller fragments of the IgG molecule were prepared and labelled. These were the $F(ab')_2$ fragment (with F_c removed by pepsin digestion) and the $F_{(ab)}$ fragment (prepared by papain digestion). All three species were investigated using time-resolved fluorescence depolarization measurements. The time-dependence of the emission anisotropies for all three species are shown in Fig. 50. The semi-loga-rithmic plot of $r(t)$ vs t is clearly non-linear for the hapten-IgG complex indicating that the molecule is not a rigid sphere. One possibility was that the molecule is a rigid ellipsoid of rotation. This could be discounted, however, since attempts to de-scribe the experimentally derived $r(t)$ in terms of a theoretically predicted anisotropy, using the results from Table 13 were unsuccessful. Thus the alternative explanation, involving some local flexibility seemed more likely. The anisotropy decay could be well described by a dual exponential function with decay times of 33 ns and 168 ns. The longer of these corresponded to the rotation of the whole body, the shorter to the flexing motion.

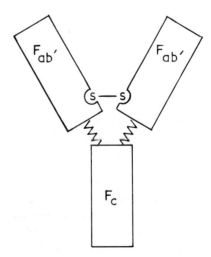

Fig. 49. Structure of Immunoglobulin G (after Valentine and Green[193])

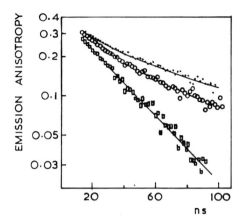

Fig. 50. Time-dependence of the emission anisotropy of dansyl-Lys bound to IgG ($\bullet$), $F_{(ab')_2}$ ($\circ$), and F_{ab} ($\square$). The *solid lines* are the least-square fit of the observed data to a single exponential decay for F_{ab}, and to a sum of two exponential decays for IgG and $F_{(ab')_2}$. (After Yguerabide et al.[222])

To determine the site of the flexibility, the two fragments were investigated. For the F_{ab} fragment the emission anisotropy decayed exponentially and the conclusion was drawn that the hapten-F_{ab} complex rotated as an essentially rigid unit. The plot of anisotropy for the hapten-$F(ab')_2$ fragment, however, shows that the decay is not due to a rigid rotor and the conclusion was drawn that the flexible hinge was sited at the junction of the two F_{ab} segments. Estimation of the angle over which the F_{ab} segments were free to rotate with respect to the whole IgG molecule yielded a value of 33 degrees. The role of the interheavy chain disulphide bond in the control of this flexibility has been reported by Chan and Cathou[19]. Comparison of the decays of emission anisotropy for native and reduced rabbit IgG showed that reduction of this bond enhanced the internal flexibility.

A study of the time-resolved depolarization of the fluorescence from a dansyl (XIII) conjugate of an IgA myeloma protein revealed no internal flexibility[219]. Comparison of these results with those for IgG, however, is not realistic since each of the several classes of immunoglobuline possess characteristic structural and biological properties.

A similar investigation of immunoglobulin M from horse, pig and nurse shark has been reported[85]. In both equine and porcine IgM extensive flexibility was observed corresponding to either

(i) hindered rotation of the F_{ab} segment or

(ii) a combination of rotations of the F_{ab} segments together with overall bending of the $F_{(ab')_2}$ fragment.

The flexibility was more restricted in the nurse shark IgM. The above results, together with a steady-state fluorescence depolarization study on the flexibility of immunoglobulins from amphibia and reptiles[226] has suggested that the degree of flexibility decreases with the level of phylogeny. Thus, time-resolved fluorescence depolarization measurements may provide one of the first techniques capable of indicating directed selection in the course of molecular evolution. Segmental flexibility has also been observed, using time-resolved fluorescence depolarization measurements, in Myosin[126].

4. Ordered Macromolecular Systems

Steady-state depolarization measurements of the fluorescence of probe molecules dispersed in the hydrocarbon interior of a lipid bilayer have been used extensively as a probe of the lipid microviscosity. This application has been reviewed[170]. Two inherent assumptions in the method, however, cast some doubt on the results obtained. Firstly the probe molecule is assumed to be a rigid spherical body rotating isotropically, a situation unlikely considering the heterogeneous nature of the system. Secondly, it is assumed that the hydrocarbon interior can be modelled using homogeneous solvents. These problems can be largely overcome with the use of time-resolved fluorescence depolarization measurements.

A popular probe molecule which has been employed for such studies is 1,6 diphenyl 1,3,5 hexatriene (DPH). This molecule has both absorption and emission along the long molecular axis and is thought to dissolve in the hydrocarbon interior. Time-resolved fluorescence depolarization studies with DPH probe molecules have been performed on the following bilayer systems: di-(dihydrosteraculoyl) phosphalidyl choline[200], dipalmitoyl phosphatidyl choline[101], L-α-dimyristollecithin residues[20], egg lecithin residues[24] and mouse leukaemic L 1210 cells[169]. In all reports the time-dependence of the emission anisotropy was found to decay non-exponentially indicating either

(i)   anisotropic rotation of the probe molecule about the axes parallel and perpendicular to the emission dipole, or

(ii) the possibility of microheterogeneity of probe sites.

Both of these invalidate the application of the steady-state technique. Dale et al.[24] found that, although the anisotropy decayed to zero for rotation of DPH in long chain paraffin solutions, in egg lecithin residues the limiting value, r_∞ was finite, an indication of the anisotropic nature of the bilayer. An investigation of the lipid L-α-dimyristollecithin in both the gel state (temperature 14.8 °C) and the fluid state (37.2 °C) has been carried out[20]. In the gel state, once again the anisotropy decayed to a limiting, finite value and the conclusion was reached that the DPH molecule was free to rotate in a cone (half angle ~30 degrees) oriented parallel to the hydrocarbon "tails" of the lipid molecules. This restriction on motion was lost with the transition to the fluid state, the decay of r(t) remaining complex but the limiting value, r_∞, tending to zero. The complex decay was explained on the basis of a microheterogeneity of DPH sites in this particular membrane.

These results are in good agreement with those measured by Lakowicz and Prendergast[116] using the differentially polarized phase fluorimetric method discussed earlier.

The influence of cholesterol on the fluidity of lipid bilayers is well known[163] and has been investigated using time-resolved fluorescence depolarization[200]. The decays of the emission anisotropy recorded for the probe molecule DPH dispersed in liposome of di-(dihydrosterculoyl) phosphalidyl choline with various added amounts

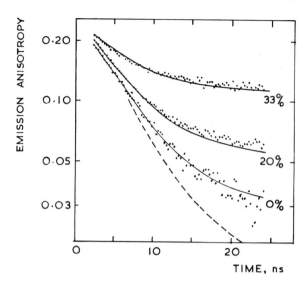

Fig. 51. Nanosecond emission anisotropy kinetics of diphenylhexatriene in liposomes consisting of di-(dihydrosterculoyl) phosphatidyl choline and varying amounts of cholesterol (0.20 and, 33 mol%). The logarithm of the emission anisotropy is plotted as a function of time. The *filled circles* represent the observed emission anisotropies. The calculated best fits of the observed data to a two-component emission anisotropy decay are shown as *solid lines*. The *broken curve* is the calculated emission anisotropy for isotropic rotation (τ_r = 6 ns). The *calculated curves* are convolutions which take into account the finite duration of the light pulse. (After Veach and Stryer[200])

of cholesterol, are displayed in Fig. 51. As may be seen, addition of cholesterol significantly reduces the rate of decay of the emission anisotropy, indicating an increase in the lipid microviscosity. Similar results have been found by Vanderkooi et al.[199] using a different fluorescent probe (12-(9-anthroyl)stearic acid).

Wahl et al.[205] have found, by time resolved fluorescence depolarization studies, that the rotational motions of proteins embedded in a membrane are greatly restricted. The fluidization of the lipid on the addition of Triton X-100 dramatically increased the decay of r(t). An obvious model system for the study of rotational motion in membranes and lipids is a probe molecule dispersed in an oriented liquid crystal. To date, however, we are aware of only one study of this type[236]. DPH was investigated in methyl cyclohexane, paraffin oil and a nematic liquid crystal aligned with an applied electric field. For the small molecule solvent, the depolarization was rapid and total indicating fast molecular reorientation. In paraffin oil the decay of the anisotropy occurred on the same time-scale as the fluorescence and the DPH molecule could be well described by a rotating prolate ellipsoid. The decays for the parallel ($I_\parallel(t)$) and perpendicular ($I_\perp(t)$) components of fluorescence for DPH dispersed in a nematic liquid crystal are shown in Fig. 52. As may be seen, the emission anisotropy does not decay during the fluorescence lifetime, indicating that the probe molecule is unable to rotate in this ordered system. The limited application of fluorescence depolarization to liquid crystalline systems is more than likely due to the lack of a clear theoretical treatment of the emission anisotropy in this case. This problem could be solved by the theoretical approach of Zannoni[225].

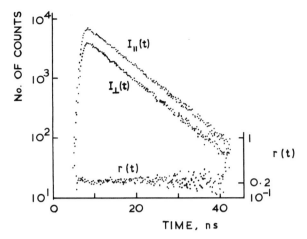

Fig. 52. Parallel ($I_\parallel$ (t)) and perpendicular ($I_\perp$ (t)) polarized components of the fluorescence decay for DPH dispersed in a nematic liquid crystal. The emission anisotropy r (t) is also shown. (After Cehelnik et al.[236])

III. Conclusions on Depolarisation Studies

In conclusion we may say that the time-resolved fluorescence depolarization experiment provides a convenient and accurate method for characterising complex molecular motion occurring in macromolecules. The limited use of the technique to date has been due to the lack of high quality fluorescence decay time data. Improvements in experimental methods associated with the use of better excitation sources such as pulsed lasers or light from synchrotron storage rings may well provide a renaissance in the use of the technique.

F. General Conclusions

In this review we have outlined experimental methods available for the study of the decay characteristic of fluorescence on a nanosecond and sub-nanosecond time-scale. These time-resolved methods have been available for a comparatively short period of time, and are improving with the provision of superior light sources. The time-scale referred to above is one which coincides with that for various molecular motions and other processes in macromolecular systems, of both synthetic and biological origin. This permits the use of fluorescence decay characteristics as a probe of such events. It is hoped that in reviewing the applications of these methods to macromolecular systems we will have stimulated the interest of polymer chemists in their use, notwithstanding the difficulties in interpretation of results which has been emphasised.

G. References

1. Adbul-Rasoul, F., et al.: Eur. Polym. J. *13*, 1019 (1977)
2. Anderson, R. A., Reid, R. F., Soutar I.: Eur. Polym. J. *15*, 925 (1979)
3. Anderson, R. M., Reid, R. F., Soutar, I.: Eur. Polym. J. *16*, 945 (1980)
4. Anufrieva, E. V., Gotlib, Y. Y.: Adv. Polym. Sci. *40*, (1981)
5. Aspler, J. S., Guillett, J. E.: Macromolecules *12*, 1082 (1979)
6. Basile, L. J.: J. Chem. Phys. *36*, 2204 (1962)
7. Beavan, S. W., Phillips, D.: Eur. Polym. J. *13*, 825 (1977)
8. Beavan, S. W., Hargreaves, J. S., Phillips, D.: Adv. Photochem. *11*, 207 (1979)
9. Beddard, G. S., Fleming, G. R., Porter, G., Robbins, R. J.: Time-Resolved Fluorescence from Biological Systems – Tryptophan and Simple Peptides. Phil. Trans. Royal Soc. 1980. In press
10. Beddard, G. S., et al.: Springer Series in Chem. Phys. *4*, 149 (1978)
11. Belford, G. G., Belford, R. L., Weber, G.: Proc. Natl. Acad. Sci. USA *69*, 1392 (1972)
12. Bevington, P. R.: Data Reduction and Error Analysis for the Physical Sciences. New York: McGraw-Hill 1969
13. Birch, D. J. S., Imhof, R. E.: J. Phys. E.: Sci. Instrum *10*, 1044 (1977)
14. Birks, J. B., Munro, I. H.: Progress in Reaction Kinetics *4*, 239 (1967)
15. Birks, J. B.: Photophysics of Aromatic Molecules. London: Wiley Interscience 1970
16. Bokobza, L., Jasse, B., Monnerie, L.: Eur. Polym. J. *13*, 921 (1977)
17. Britten, A., Lockwood, G.: Mol. Photochem. *7*, 79 (1976)
18. Chakrabarti, S. K., Ware, W. R.: J. Chem. Phys. *55*, 5494 (1971)
19. Chan, L. M., Cathou, R. E.: J. Mol. Biol. *112*, 653 (1977)
20. Chen, L. A. et al.: J. Biol. Chem. *252*, 2163 (1977)
21. Costa, S. M. de. B. et al.: J. Photochem. *12*, 11 (1979)
22. Cummins, H. Z. et al.: J. Chem. Phys. *9*, 518 (1969)
23. Cundall, R. B., Evans, G. B.: J. Phys. E.: Sci. Instrum *1*, 305 (1968)
24. Dale, R. E., Chen, L. A., Brand, L.: J. Biol. Chem. *252*, 7500 (1977)
25. Dandliker, W. B., Portmann, A. J.: Fluorescent Protein Conjugates. In: Excited States of Proteins and Nucleic Acids. Steiner, R. F., Weinigh, I. (Eds.) p. 199, New York: Plenum 1971
26. David, C., Lempereur, M., Gueskens, G.: Eur. Polym. J. *8*, 417 (1972)
27. David, C., Lempereur, M., Gueskens, G.: Eur. Polym. J. *9*, 1315 (1973)
28. David, C., Putman de Lavaraille, N., Gueskens, G.: Eur. Polym. J. *10*, 617 (1974)
29. David, C., Lempereur, M., Gueskens, G.: Eur. Polym. J. *10*, 1181 (1974)
30. David, C., Piens, M., Gueskens, G.: Eur. Polym. J. *12*, 621 (1976)
31. David, C., Baeyens-Volant, D., Piens, M.: Eur. Polym. J. *16*, 413 (1980)
32. Demas, J. N., Crosby, G. A.: J. Phys. Chem. *75*, 991 (1971)
33. De Toma, R. P., Easter, J. H., Brand, L.: J. Am. Chem. Soc. *98*, 5001 (1976)
34. Dubois-Violette, E. et al.: J. Chim. Phys. *66*, 1865 (1969)
35. Durbin, J., Watson, G. S.: Biometrika *37*, 409 (1950)
36. Durbin, J., Watson, G. S.: Biometrika *38*, 159 (1951)
37. Dwek, R. A.: NMR in Biochemistry. Clarendon Press Oxford, U.K. 1973
38. Easter, J. H., De Toma, R. P., Brand, L.: Biophys. J. *16*, 571 (1976)
39. Easter, J. H., De Toma, R. P., Brand, L.: Biochem. Biophys. Acta *508*, 27 (1978)
40. Edelman, G. M., McClure, W. O.: Accounts Chem. Res *1*, 65 (1968)
41. Eisinger, J., Lamola, A. A.: The excited states of nucleic acids. In: Excited States of Proteins and Nucleic Acids. Steiner, R. F., Weinryl, I. (Eds.) p. 107, New York: Plenum 1971
42. Feinstein, A., Rowe, A. J.: Nature *205*, 147 (1965)
43. Fleming, G. R., Morris, J. M., Robinson, G. W.: Chem. Phys. *17*, 91 (1976)
44. Fleming, G. R., Morris, J. M., Robinson, G. W.: Austral. J. Chem. *30*, 2337 (1977)
45. Fleming, G. R. et al.: Chem. Phys. Letts. *52*, 228 (1977)
46. Fleming, G. R. et al.: Chem. Phys. Letts. *49*, (1977)

47. Fleming, G. R., Beddard, G. J.: CW mode-locked dye lasers for ultra fast spectroscopic studies. In: Optics and Laser Technology. p. 257 (1978)
48. Fleming, G. R. et al.: Proc. Natl. Acad. Sci. USA 75, 4652 (1978)
49. Fleming, G. R., Knight, A. E. W., Morris, J. M., Morrison, R. J. S., Robbins, R. J., Robinson, G. W.: Picosecond Fluorescence Studies of the Fluorescence Probe 1,8-anilino-naphthalene sulphonate (ANS). Israel J. Chem. In press
50. Forster, Th.: Disc. Faraday Soc. 27, 1 (1939)
51. Fox, R. B. et al.: J. Chem. Phys. 57, 534 (1972)
52. Fox, R. B. et al.: J. Chem. Phys. 57, 2284 (1972)
53. Fox, R. B.: Pure Appl. Chem. 30, 87 (1972)
54. Franck, C. W., Harrah, L. A.: J. Chem. Phys. 61, 1526 (1974)
55. Gafni, A. et al.: Biophys. J. 17, 155 (1977)
56. Ghiggino, K. P., Nicholls, C. H., Pailthorpe, M. T.: J. Photochem. 4, 155 (1975)
57. Ghiggino, K. P. et al.: J. Photochem. 7, 141 (1977)
58. Ghiggino, K. P., Roberts, A. J., Phillips, D.: J. Photochem. 9, 301 (1978)
59. Ghiggino, K. P., Wright, R. D., Phillips, D.: Chem. Phys. Letts. 53, 552 (1978)
60. Ghiggino, K. P., Wright, R. D., Phillips, D.: J. Polym. Sci. (Polym. Phys.) 16, 1499 (1978)
61. Ghiggino, K. P., Wright, R. D., Phillips, D.: Eur. Polym. J. 14, 567 (1978)
62. Ghiggino, K. P. et al.: J. Photochem. 11, 297 (1979)
63. Ghiggino, K. P., Roberts, A. J., Phillips, D.: Springer Series in Chem. Phys. 6, 98 (1979)
64. Ghiggino, K. P., Lee, A. G., Meech, S. R., O'Connor, D. V., Phillips, D.: Time-Resolved Emission Spectroscopy of the Dansyl Fluorescence Probe. BIOCHEM (in press)
65. Ghiggino, K. P., Roberts, A. J., Phillips, D.: J. Phys. E.: Sci. Instrum., 13, 446 (1980)
66. Goldenberg, M., Emert, J., Morawetz, H.: J. Amer. Chem. Soc. 100, 7171 (1978)
67. Gordon, R. G.: J. Chem. Phys. 45, 1643 (1966)
68. Grinwald, A., Steinberg, I. Z.: Biochem. 13, 5170 (1974)
69. Grinwald, A., Steinberg, I. Z.: Anal. Biochem. 59, 583 (1974)
70. Grinwald, A., Steinberg, I. Z.: Biochim. Biophys. Acta. 427, 663 (1976)
71. Grinwald, A., Steinberg, I. Z.: Biophys. J. 19, 74 (1977)
72. Guilbault, G. C.: Practical Fluorescence, Theory, Methods, and Techniques. New York: Dekker 1973
73. Hackett, P. A., Phillips, D.: J. Phys. Chem. 78, 671 (1974)
74. Hara, K. et al.: Chem. Phys. Letts. 69, 105 (1980)
75. Hare, F., Lusson, C., Sarchez, E.: J. Chim. Phys. Phys. Chem. Biol. 73, 621 (1976)
76. Harrah, L. A.: J. Chem. Phys. 56, 385 (1972)
77. Harvey, S. C., Cheung, H. C.: Proc. Natl. Acad. Sci. USA 69, 3670 (1972)
78. Heiftje, G. M., Hangen, G. R., Ramsey, J. M.: Appl. Phys. Lett. 30, 463 (1977)
79. Heisel, F., Laustriat, G.: J. Chim. Phys. 66, 1895 (1969)
80. Helene, C.: Excited state interactions and energy transfer processes in the photochemistry of protein-nucleic acid complexes. In: Excited States of Biological Molecules. Birks, J. B. (Ed.) p. 151, London – New York – Toronto – Sydney: Wiley 1976
81. Hirayama, F.: J. Chem. Phys. 42, 3163 (1965)
82. Hirayama, F., Basile, L. J., Kikuchi, C.: Mol. Cryst. 4, 83 (1968)
83. Hirayama, S., Phillips, D.: J. Photochem. 12, 139 (1980)
84. Holden, D. A., Wang, P. Y.-K., Guillet, J. E.: Macromolecules 13, 295 (1980)
85. Holowka, D. A., Cathou, R. E.: Biochem. 15, 3379 (1976)
86. Hoyle, C. E., Guillet, J. E.: Macromolecules 12, 956 (1979)
87. Hoyle, C. E. et al.: Macromolecules 11, 429 (1978)
88. Irie, M. et al.: J. Phys. Chem. 81, 1571 (1977)
89. Isenberg, I., Dyson, R. D.: Biophys. J. 9, 1337 (1969)
90. Isenberg, I.: J. Chem. Phys. 59, 5696 (1973)
91. Ishi, T., Handa, T., Matsunaga, S.: Makromol. Chem. 178, 2357 (1977)
92. Ishi, T., Handa, T., Matsunaga, S.: Macromolecules 11, 40 (1978)
93. Itaya, A., Okamoto, K., Kusabayashi, S.: Bull. Chem. Soc. Japan 49 II, 2082 (1976)
94. Ito, S., Yamamoto, M., Nishijima, Y.: Rep. Prog. Polym. Phys. Japan. 19, 421 (1976)

95. Ito, S., Yamamoto, M., Nishijima, Y.: Repts. Prog. Polym. Phys. Japan *20*, 481 (1977)
96. Jablonski, A.: Bull. Acad. Polon. Sci., Ser. Sci. Math., Astr. Phys. *8*, 259 (1960)
97. Jablonski, A.: Z. Naturforsch. *164*, 1 (1961)
98. Jablonski, A.: Decay and polarization of fluorescence of solutions. In: Luminescence of Organic and Inorganic Materials. H. P. Kallmann, G. M. Spruch (Eds.) p. 110, Wiley 1962
99. Jarry, J. P., Monnerie, L.: J. Polym. Sci. Polym. Phys. E. *16*, 443 (1978)
100. Johnson, G. E.: J. Chem. Phys. *62*, 4697 (1975)
101. Kawato, S., Kinosita, K., Ikegami, A.: Biochem. *16*, 2319 (1977)
102. Keyanpour-Rad, M., Ledwith, A., Johnson, G. E.: Macromolecules *13*, 222 (1980)
103. Kinosita, K., Kawato, S., Ikegami, A.: Biophys. J. *20*, 289 (1977)
104. Kirby, E. P.: Fluorescence instrumentation and methodology. In: Excited States of Proteins and Nucleic Acids. Steiner, R. F., Weinryb, I. (Eds.) p. 31, New York: Plenum 1971
105. Kirkov-Kaminska, E., Rothiewicz, K., Gradbowska, A.: Chem. Phys. Letts. *58*, 379 (1978)
106. Klöpffer, W.: J. Chem. Phys. *50*, 2337 (1969)
107. Klöpffer, W.: Intramolecular excimers in: Organic Molecular Photophysics, Vol. I. J. B. Birks (Ed.) London: Wiley Interscience 1973
108. Knight, A. G., Selinger, B. K.: Spectrochim. Acta *27A*, 1223 (1971)
109. Knight, A. E. W., Selinger, B. K.: Austral. J. Chem. *26*, (1973)
110. Koester, V. J., Dowben, R. M.: Rev. Sci. Instrum. *49*, 1186 (1978)
111. Konev, S. V.: Fluorescence and Phosphorescence of Proteins and Nucleic Acids. New York: Plenum 1967
112. Kosower, E. M., Doduik, H.: J. Am. Chem. Soc. *96*, 6195 (1974)
113. Kosower, E. M. et al.: J. Am. Chem. Soc. *97*, 2167 (1975)
114. Kuntz, E.: Phosphorescence instrumentation and techniques. In: Excited States of Proteins and Nucleic Acids, Steiner, R. F., Weinryb, I. (Eds.) p. 87, New York: Plenum 1971
115. Lakowicz, J. R., Weber, G.: Biochem. *12*, 4171 (1973)
116. Lakowicz, J. R. Prendergast, F. G.: Science *200*, 1399 (1978)
117. Laws, W. R., Brand, L.: J. Phys. Chem. *83*, 782 (1979)
118. Lee, A. G. et al.: FEBS Lett. *94*, 17 (1978)
119. Lee, A. G., Green, N. M., Ghiggino, K. P., Phillips, D.: Unpublished results
120. Lee, T. Y., Fugate, R. D.: Photochem. Photobiol. *27*, 803 (1978)
121. Lehrer, S. S.: Biophys. J. *19*, (1977)
122. Lerner, J., Lami, H.: Electronic energy transfer in some class B proteins: trypsin, lysozyme, α-chymotrypsin and chymotrypsin A. In: Excited States of Biological Molecules. Birks, J. B. (Ed.) p. 601, London – New York – Toronto – Sydney: Wiley 1976
123. Lombardi, J. R., Dafforn, G. A.: J. Chem. Phys. *44*, 3882
124. Meech, S. R., Phillips, D.: Time-Resolved Fluorescence Spectroscopy of Tryptophan Analogues in Lipid Bilayers. Unpublished results 1980
125. Memming, R.: Z. Phys. Chem. *28*, 168 (1961)
126. Mendelson, R. A., Morales, M. F., Botts, J.: Biochem. *12*, 2250 (1973)
127. Menzel, E., Popovic, Z. D.: Rev. Sci. Instrum. *49*, 39 (1978)
128. Meserve, E. T.: Direct measurement of fluorescence lifetimes. In: Excited States of Proteins and Nucleic Acids. Steiner, R. F., Weinryb, I. (Eds.) p. 57, New York: Plenum 1971
129. Morawetz, H.: Fluorescence studies of conformational mobility and of the mutual interpenetration of flexible chain molecules. WPAC Symposium Macromol. Chem. Mainz. In press, communicated privately
130. Muller, A., Lumry, R., Kokubun, H.: Rev. Sci. Instrum. *36*, 1214 (1965)
131. Nishihara, T., Kaneko, M.: Makromol. Chem. *124*, 84 (1969)
132. Nishijima, Y.: Fluorescence methods in polymer research. In: Progress in Polymer Science, Japan, *6*. Onogi, S., Uno, K. (Eds.) p. 199, New York: John Wiley 1973
133. Nishijima, Y., Yamamoto, M.: Polymer Reprints *20*, 391 (1979)
134. Noelken, M. E. et al.: J. Biol. Chem. *240*, 218 (1965)
135. North, A. M., Soutar, I.: J.C.S. Faraday I *68*, 1101 (1972)
136. O'Connor, D. V., Ware, W. R., Andre, J. C.: J. Phys. Chem. *83*, 1333 (1979)

137. O'Konski, C. T.: Electric birefringence and relaxation in solution of rigid macromolecules. In: Molecular electro-optics, Part 1, Theory and Methods. C. T. O'Konski (Ed.) p. 68, New York: Marcel Dekker 1976
138. Parker, C. A.: Photoluminescence of Solutions. Jennings, K. R., Cundall, R. B. (Eds.) Amsterdam: Elsevier 1968
139. Pailthorpe, M. T.: J. Phys. E. Sci. Instrum. *8*, 194 (1975)
140. Paulson, C. M.: Electric dichroism of macromolecules. In: Molecular Electrooptic, Part 1, Theory and Methods. C. T. O'Korski (Ed.) p. 63, New York: M. Dekker, Inc. 1976
141. Pecorra, R.: J. Chem. Phys. *40*, 1604 (1964)
142. Perrin, F.: J. Phys. Radium *5*, 497 (1934)
143. Perrin, F.: J. Phys. Radium *7*, 1 (1936)
144. Phillips, D., Roberts, A. J., Soutar, I.: Transient decay studies of photophysical processes in aromatic ploymers, I. Multiexponential fluorescence decays in copolymers of 1-vinyl naphthalene and methyl methacrylate. J. Polymer. Sci. Polym. Phys. Ed. *18* (No. 12) (1980)
145. Phillips, D., Roberts, A. J., Soutar, I.: Transient decay studies of photophysical processes in aromatic polymers, II. Investigation of intramolecular excimer formation in copolymers of 1-vinyl naphthalene and methyl acrylate. Polymer *22*, 293 (1981)
146. Phillips, D., Roberts, A. J., Soutar, I.: Transient decay studies of photophysical processes in aromatic polymers, III. Concentration dependence of excimer formation in copolymers of acenaphthylene and methyl methacrylate. Eur. Polym. J. *17*, 101 (1981)
147. Phillips, D., Roberts, A. J., Soutar, I.: Transient decay studies of photophysical processes in aromatic polymers. V Temperature dependence of excimer formation and decay in copolymers of 1-vinyl naphthalene. As yet unpublished 1980
148. Phillips, C. F.: Modulation techniques in chemical kinetics. In: Progress in Reaction Kinetics *7*, p. 71, New York: Pergamon 1973
149. Porter, G., Sadkowski, P. J., Treadwell, C. J.: Chem. Phys. Letts. *49*, 416 (1977)
150. Rayner, D. M. et al.: Can. J. Chem. *54*, 3246 (1976)
151. Rayner, D. M., Krajcarski, D. T., Szabo, A. G.: Can. J. Chem. *56*, 1238 (1978)
152. Rayner, D. M., Szabo, A. G.: Can. J. Chem. *56*, 743 (1978)
153. Reeves, R. L., Maggio, M. S., Costa, C. F.: J. Am. Chem. Soc. *96*, 5917 (1974)
154. Reid, R. F., Soutar, I.: J. Polym. Sci. Polym. Letts. *15*, 153 (1977)
155. Reid, R. F., Soutar, I.: J. Polym. Sci. Polym. Phys. *16*, 231 (1978)
156. Reid, R. F., Soutar, I.: J. Polym. Sci. Polym. Phys. *18*, 457 (1980)
157. Roberts, A. J., Cureton, C. G., Phillips, D.: Chem. Phys. Letts. *72*, 554 (1980)
158. Roberts, A. J., O'Connor, D. V., Phillips, D.: Multicomponent fluorescence decay in vinyl aromatic polymers and copolymers. Annal. N.Y. Acad. Sci. *366*, 109 (1981)
159. Roberts, A. J., Phillips, D., Abdul-Rasoul, F., Ledwith, A.: Temperature dependence of excimer formation and dissociation of poly(N-vinyl carbazole). J.C.S. Faraday I. (in press) 1980
160. Robinson, G. W., Caughey, T. A., Auerbach, R. A.: Picosecond emission spectroscopy with an ultraviolet sensitive streak camera. In: Advances in Laser Chemistry/Springer Series in Chemical Physics. Zewail, A. H. (Ed.) p. 108, Berlin, Heidelberg, New York: Springer 1978
161. Robinson, G. W. et al.: J. Mol. Struct. *47*, 221 (1978)
162. Roe, R. J.: J. Polym. Sci. Polym. Phys. Ed. *8*, 1187 (1970)
163. Rogers, J., Lee A. G., Wilton, D. C.: Biochem. Biophys. Acta *552*, 23 (1979)
164. Rothkiewicz, K., Grabowski, A. R., Jasny, J.: Chem. Phys. Letts. *34*, 55 (1975)
165. Rothschild, W. G., Rosasco, G. J., Livingston, R. C.: J. Chem. Phys. *62*, 1253 (1975)
166. Sackmann, E.: Z. Physikal. Chem. *101*, 391 (1976)
167. De Schryver, F. C., Demeyer, K., van der Auweraer, M., Quanten, E.: Excimer Formation in Poly- and Di-Chromophoric Molecules in Solution. Annal. N.Y. Acad. Sci. In press, communicated privately 1980
168. Seliskar, C. J., Brand, L.: J. Am. Chem. Soc. *93*, 5405 (1971)
169. Sene, C. et al.: FEBS Letts. *88*, 181 (1978)
170. Shinitzky, M., Barenholtz, Y.: Biochem. Biophys. Acta. *515*, 367 (1978)

171. Sloman, A. W., Swords, M. D.: J. Phys. E.: Sci. Instrum. *11*, 521 (1978)
172. Small, E. W., Isenberg, I.: Biopolym. *16*, 1907 (1977)
173. Somersall, A. C., Guillet, J. E.: Macromolecules *6*, 218 (1973)
174. Somersall, A. C., Guillet, J. E.: J. Macromol. Sci. Rev. Macromol. Chem. *13C*, 135 (1975)
175. Spears, K. G., Rice, S. A.: J. Chem. Phys. *53*, 5561 (1971)
176. Spears, K. G., Cramer, L. E., Hoffland, L. D.: Rev. Sci. Instrum. *49*, 255 (1978)
177. Spencer, R. D., Weber, G.: Ann. N.Y. Acad. Sci. *158*, 361 (1969)
178. Spencer, R. D., Weber, G.: J. Chem. Phys. *52*, 1654 (1970)
179. Steiner, R. F., Weinryb, I. (Eds.): Excited States of Proteins and Nucleic Acids. New York: Plenum 1971
180. Strickler, S. J., Berg, R. A.: J. Chem. Phys. *37*, 814 (1962)
181. Stryer, L.: Science *162*, 526 (1968)
182. Stryer, L., Griffith, O. H.: Proc. Natl. Acad. Sci. USA *54*, 1785 (1965)
183. Szabo, A. G., Rayner, D. M.: Biochem. Biophys. Res. Communications *94*, 909 (1980)
184. Tagawa, S., Washio, M., Tabata, Y.: Chem. Phys. Letts. *68*, 276 (1979)
185. Takashima, S.: Dielectric properties of proteins, I. Dielectric relaxation. In: Physical Principles and Techniques of Protein Chemistry, Part A. Leach, S. J. (Ed) p. 291, New York: Academic Press Inc. 1969
186. Tanford, C.: Physical Chemistry of Macromolecules. New York: John Wiley & Sons. Inc. p. 437, 1961
187. Tao, T.: Biopolym. *8*, 609 (1969)
188. Teale, F. W. J., Weber, G.: Biochem. J. *65*, 476 (1957)
189. Thomas, D. D.: Biophys. J. *24*, 439 (1978)
190. Thulborn, K. R., Sawyer, W. H.: Biochem. Biophys. Acta *511*, 125 (1978)
191. Tschanz, H. P., Binkert, Th.: J. Phys. E.: Sci. Instrum. *9*, 1131 (1976)
192. Vala, M. J. jr., Haebig, J., Rice, S. A.: J. Chem. Phys. *43*, 886 (1965)
193. Valentine, R. C., Green, N. M.: J. Mol. Biol. *27*, 615 (1967)
194. Valeur, B., Monnerie, L.: C. R. Acad. Sci. Paris *280C*, 57 (1975)
195. Valeur, B. et al.: J. Polym. Sci., Polym. Phys. Ed. *13*, 667 (1975)
196. Valeur, B., Monnerie, L., Jarry, J. P.: J. Polym. Sci., Polym. Phys. Ed. *13*, 675 (1975)
197. Valeur, B., Monnerie, L.: J. Polym. Sci., Polym. Phys. E. *14*, 11 (1976)
198. Valeur, B., Monnerie, L.: J. Polym. Sci., Polym. Phys. E. *14*, 29 (1976)
199. Vanderkooi, J. et al.: Biochem. *13*, 1589 (1974)
200. Veach, W. R., Stryer, L.: J. Mol. Biol. *117*, 1109 (1977)
201. Wahl, P.: Compt. Rend. Acad. Sci. Paris *260*, 6891 (1965)
202. Wahl, P.: Compt. Rend. Acad. Sci. *263D*, 1525 (1966)
203. Wahl, P., Timasheff, S. N.: Biochem. *8*, 2945 (1969)
204. Wahl, P., Meyer, G., Parrod, J.: Europ. Polym. J. *6*, 585 (1970)
205. Wahl, P. et al.: Eur. J. Biochem. *18*, 332 (1971)
206. Wahl, P., Brochon, J. C.: Resolution of the Fluorescence Spectra of Proteins Using Decay Measurements. Dyn. Aspects Conform. Changes Biol. Macromol. Proc. Annu. Meet. Soc. Chim. Phys. 23rd. Sadron, C. (Ed.) p. 129, Dardrecht: Meth. Maidel 1973
207. Wahl, P., Auchet, J. C., Danzel, B.: Rev. Sci. I. *45*, 28 (1974)
208. Wahl, P.: Chem. Phys. *7*, 210 (1975)
209. Wahl, P.: Chem. Phys. *22*, 245 (1977)
210. Wang, Y. C., Morawetz, H.: Makromol. Chem. Suppl. *1*, 283 (1975)
211. Ware, W. R.: Transient luminescence measurements. In: Creation and Detection of the Excited State. Lamola, A. A. (Ed.) p. 213, *1A*, New York: Marcel-Dekker 1971
212. Ware, W. R. et al.: J. Chem. Phys. *54*, 4729 (1971)
213. Ware, W. R., Doemeny, L. J., Nemzek, T. L.: J. Phys. Chem. *77*, 2038 (1973)
214. Weber, G.: J. Chem. Phys. *55*, 2399 (1971)
215. Weber, G.: What we have learnt about proteins from the study of their photoexcited states. In: Excited States of Biological Molecules. Birks, J. B. (Ed.) p. 363, London – New York – Toronto – Sydney: Wiley 1976
216. Weber, G. et al.: Biochem. *15*, 4429 (1976)

217. Weber, G.: J. Chem. Phys. *69*, 4081 (1977)
218. Weber, G.: Resolution of the fluorescence lifetimes in a heterogeneous system by phase and modulation measurements. J. Chem. Phys. In press, communicated privately
219. Weltman, J. K., Davis, R. P.: J. Mol. Biol. *54*, 177 (1970)
220. Williams, G.: Chem. Soc. Rev. *7*, 89 (1978)
221. Yanari, S. S., Bovey, F. A., Lumry, R.: Nature *200*, 242 (1963)
222. Ygeurabide, J., Epstein, H. F., Stryer, L.: J. Mol. Biol. *51*, 573 (1970)
223. Ygeurabide, J.: Nanosecond fluorescence spectroscopy of macromolecules. In: Methods in Enzymology. Hirs, C. H. W., Timasheff, S. N. (Eds.) p. 498, New York: Academic Press 1972
224. Ygeurabide, J.: In: Fluorescence Techniques. Thaer, A. A., Sernetz, M. (Eds.) Berlin, Heidelberg, New York: Springer Verlag 1973
225. Zannoni, C.: Mol. Phys. *38*, 1813 (1979)
226. Zagyansky, Y. A.: Arch. Biochem. Biophys. *166*, 371 (1975)
227. Zinzli, P. E.: J. Phys. E.: Sci. Instrum. *11*, 17 (1978)
228. Longworth, J. W.: Luminescence of polypeptides and proteins, In: Excited States of Proteins and Nucleic Acids. Steiner, R. F., Weinryb, I. (Eds.) p. 319, New York: Plenum 1971
229. McKinnon, A. E., Szabo, A. G., Miller, D. R.: J. Phys. Chem. *81*, 1564 (1977)
230. Block, H., North, A. M.: Adv. Mol. Relax Process. *1*, 309 (1970)
231. Burstein, E. A., Vedenkin, N. S., Ivkova, M. N.: Photochem. Photobiol. *18*, 263 (1973)
232. Mantolin, W. W., Pownall, H. J.: Photochem. Photobiol. *26*, 69 (1977)
233. Weber, G.: Biochem. J. *75*, 335 (1960)
235. Beddard, G. S. et al.: Phil. Trans. R. Soc. Lond. A. *298*, 321 (1980)
236. Cehelnik, E. D. et al.: J. Chem. Soc. Faraday II, *10*, 244 (1974)
237. Irwin, D. J., Livingstone, A. E.: Computer Phys. Commun. *7*, 95 (1974)

Received February 3, 1981
M. Gordon (editor)

Author Index Volumes 1–40